Management and Industrial Engineering

Series Editor

J. Paulo Davim, Department of Mechanical Engineering, University of Aveiro, Aveiro, Portugal

This series fosters information exchange and discussion on management and industrial engineering and related aspects, namely global management, organizational development and change, strategic management, lean production, performance management, production management, quality engineering, maintenance management, productivity improvement, materials management, human resource management, workforce behavior, innovation and change, technological and organizational flexibility, self-directed work teams, knowledge management, organizational learning, learning organizations, entrepreneurship, sustainable management, etc. The series provides discussion and the exchange of information on principles, strategies, models, techniques, methodologies and applications of management and industrial engineering in the field of the different types of organizational activities. It aims to communicate the latest developments and thinking in what concerns the latest research activity relating to new organizational challenges and changes world-wide. Contributions to this book series are welcome on all subjects related with management and industrial engineering. To submit a proposal or request further information, please contact Professor J. Paulo Davim, Book Series Editor, pdavim@ua.pt.

Carolina Machado
Editor

Ethics in Management and Business

Springer

Editor
Carolina Machado
Department of Management, School of Economics, Management and Political Science
University of Minho
Braga, Portugal

ISSN 2365-0532 ISSN 2365-0540 (electronic)
Management and Industrial Engineering
ISBN 978-3-031-82053-3 ISBN 978-3-031-82054-0 (eBook)
https://doi.org/10.1007/978-3-031-82054-0

This Springer imprint is published by the registered company Springer Nature Switzerland AG
The registered company address is: Gewerbestrasse 11, 6330 Cham, Switzerland

Preface

Ethics plays a fundamental role in management and business, laying the foundation for responsible and sustainable business practices. In a dynamic and interconnected business environment, integrity in decision-making and daily operations not only builds a positive reputation but also contributes to long-term value creation. More than a peripheral consideration, ethics is the driving force behind business success that transcends finances, incorporating moral values and social responsibility. In management, ethical decision-making is at the heart of responsible leadership. This goes beyond the pursuit of profit; it's about considering the impact of each decision on stakeholders and society as a whole. Corporate Social Responsibility (CSR) is not a mere formality but an active commitment to contributing to social and environmental well-being. CSR transcends mere compliance with regulations. Ethical companies go further, adopting practices that benefit society and the environment. This includes sustainability initiatives, supporting local communities, and ethical work practices. Transparency and accountability form the foundation of business ethics. Accountability creates an environment in which organizations are responsible for their actions. Open communication about practices, results, and challenges not only builds trust but also drives an ethical organizational culture. An ethical culture permeates all levels of the organization. Leaders play a central role in this culture, setting the tone and influencing integrity at all company levels. Diversity and inclusion are not only buzzwords but also ethical principles that promote an equitable and innovative work environment. Ethical companies value diversity of perspectives, recognizing that it drives innovation and promotes a more equitable workplace. Compliance is not limited to laws and regulations but is reinforced by strong governance systems, including codes of conduct and internal audits. Ethics extends beyond company borders to supply chains. Ethical companies ensure that their partners and suppliers share similar values, avoiding exploitative practices. From this ethical perspective, innovation does not only focus on commercial benefits but also carefully evaluates social and environmental impacts. This involves considering the ethical implications when developing new products or services. In the digital universe, ethics are crucial for the responsible management of data and privacy. Ethical companies adopt transparent and safe policies, respecting the trust placed by customers. When ethical challenges arise, ethical companies meet them head-on, demonstrating not only integrity but also a constant commitment to correction and continuous improvement. In short, ethics in management and business is not an accessory. It's an essential strategy for long-term and sustainable success. Ethical companies not only thrive in the present but also contribute to a more equitable, transparent, and socially responsible business future.

Conscious of these challenges, with this book, entitled *Ethics in Management and Business*, we look to present the different contributions that have been made for a deeper understanding of the challenges, opportunities, and best practices related to ethics in management and business. It looks to provide managers, leaders, as well as all kinds of professionals with the most recent and critical knowledge and abilities/competencies, to deal with the processes, tools, and mechanisms able to build ethical and successful organizations. *Ethics in Management and Business* will make a huge contribution to organizations assuming their responsibility toward the environment, promoting sustainable business practices, and minimizing environmental impact, thus achieving concrete and competitive results and being prepared for new and continuous challenges. To this end, different experts from different fields of research were invited, throughout the various chapters that constitute this book, to report on the strategies, models, practices, and tools able to provide the executives, managers, and engineers with the knowledge and good practices needed to promote ethics in the management and business field.

Contributing to stimulating the growth and development of each individual in a dynamic, competitive, global, responsible, and sustainable economy, *Ethics in Management and Business*, organized into eight chapters, starts by discussing "Adorno, Ethics and Business Ethics." Followed by the analysis of "Confucian Diligence: A Core Ethical Value That Shapes Business Success in East Asia," the next chapter highlights "The Ethics of Digital Nudging." It follows an approach to "Corporations as Communicative Moral Agents," analyzing the following chapter, "Is Applying Ethics During a Transitional Period Ethical? A Review of Ethical Leadership and Its Influence on Employees and the Organization." ""Greasing the Wheels of Commerce" or "Sand in the Machine"? The Limits of Traditional Economic Thinking on Bribery in Business and Some Business Ethical Thoughts" is also a target of analysis, focusing on the next chapter "Ethical Leadership: Its Relationship with Social Responsibility." The book ends with a discussion about "Sexual Harassment in an Organizational Context: Ethical Decision-Making Process."

Ethics in Management and Business can be used by various potential stakeholders, including academics/researchers, managers, engineers, practitioners, and other professionals in the different areas of business, management, and engineering. Students from different undergraduate/graduate levels (undergraduate, master, and doctoral/Ph.D.s) in the areas of management and industrial engineering are also potential stakeholders in this book, who need to know and understand the challenges that organizations are facing nowadays, namely (and just for example) at the level of corporate social responsibility, ethics in leadership, ethics in negotiations and contracts, digital and technological ethics, corporate sustainability, obtaining, in this way, the knowledge necessary to build more responsible and sustainable organizations, as well as a more sustainable society, for present and future generations.

The Editor acknowledges her gratitude to Springer for this opportunity and their professional support. Finally, we would like to thank all chapter authors for their interest and availability to work on this project.

Braga, Portugal Carolina Machado

Contents

Editor and Contributors

About the Editor

Carolina Machado received her Ph.D. degree in Management Sciences (Organizational and Policies Management area/Human Resources Management) from the University of Minho in 1999, her Master's degree in Management (Strategic Human Resource Management) from the Technical University of Lisbon in 1994, and Degree in Business Administration from the University of Minho in 1989. Teaching in Human Resources Management subjects since 1989 at the University of Minho; she is since 2004 Associate Professor (with Habilitation since 2022), with experience and research interest areas in the field of Human Resource Management, International Human Resource Management, Human Resource Management in SMEs, Training and Development, Emotional Intelligence, Management Change, Knowledge Management and Management/HRM in the Digital Age/Business Analytics, Sustainability and Higher Education Sustainability. She is Head of the Human Resources Management Work Group at the School of Economics, Management and Political Science at University of Minho, Coordinator of Advanced Training Courses at the Interdisciplinary Centre of Social Sciences, Member of the Interdisciplinary Centre of Social Sciences (CICS.NOVA.UMinho), University of Minho, as well as Chief Editor of the *International Journal of Applied Management Sciences and Engineering* (IJAMSE), Guest Editor of journals, books Editor and book Series Editor, as well as reviewer in different international prestigious journals. In addition, she has also published both as editor/co-editor and as author/co-author of several books, book chapters, and articles in journals and conferences. e-mail: carolina@eeg.uminho.pt

Contributors

Florian Beham RPTU School of Business and Economics, University of Kaiserslautern-Landau, Kaiserslautern, Germany

Sophia Chia-Min Chou Graduate Institute of Management, National Taiwan Normal University, Da'an Dist., Taipei City, Taiwan

José Maria L. de Pedro Universidad Villanueva, Madrid, Spain

Ana João Gonçalves Ferreira School of Economics, Management and Political Science, University of Minho, Braga, Portugal

Florian Fuchs School of Economic Disciplines, University of Siegen, Siegen, Germany

Cristiana Cerqueira Leal Centre for Research in Economics and Management (NIPE) & School of Economics, Management and Political Science, University of Minho, Braga, Portugal

Volker Lingnau RPTU School of Business and Economics, University of Kaiserslautern-Landau, Kaiserslautern, Germany

Carolina Feliciana Machado School of Economics, Management and Political Science, University of Minho, Braga, Portugal;
Interdisciplinary Centre of Social Sciences (CICS.NOVA.UMinho), University of Minho, Braga, Portugal

Jaakko Nevasto Essex Business School, University of Essex, Colchester, UK

Benilde Oliveira School of Economics, Management and Political Science, University of Minho, Braga, Portugal

Craig Reeves School of Law, Birkbeck College, University of London, London, UK

Beatriz Sousa Silva School of Economics, Management and Political Science, University of Minho, Braga, Portugal

Matthew Sinnicks Southampton Business School, University of Southampton, Southampton, UK

Paulo Viana School of Economics, Management and Political Science, University of Minho, Braga, Portugal

Adorno, Ethics and Business Ethics

Craig Reeves, Jaakko Nevasto, and Matthew Sinnicks

Abstract Theodor W. Adorno was one of the twentieth century's most potent and influential European thinkers, whose impact is felt across the humanities and social sciences. However, Adorno's thought has been almost entirely absent from the business ethics conversation. This chapter explores the relevance of Adorno's thought for business ethics that has emerged in recent scholarship. It does so through an engagement with topics such as positivistic management, consumer culture, social media and political discourse, and the possibility of good work, and by expounding Adorno's critical theory, the critique of the business ethics tradition implied by his work, and the debate regarding whether Adorno can be read as a 'negative Aristotelian'.

Keywords Adorno · Critical theory · Business ethics · Moral philosophy · Frankfurt school

1 Introduction

Theodor W. Adorno was one of the twentieth century's most potent and influential European thinkers, whose impact is felt across the humanities and social sciences. Yet, though he has attracted some attention from Critical Management Studies scholars (Parker 2003; Beyes and De Cock 2017; Granter 2019) on account of his powerful critique of our 'wrong life', until recently Adorno's thought has been almost entirely absent from the business ethics conversation (Reeves and Sinnicks 2021, 2022). This chapter explores the relevance of Adorno's thought for business ethics that has emerged in recent scholarship.

C. Reeves
School of Law, Birkbeck College, University of London, London, UK

J. Nevasto
Essex Business School, University of Essex, Colchester, UK

M. Sinnicks (✉)
Southampton Business School, University of Southampton, Southampton, UK
e-mail: m.sinnicks@soton.ac.uk

C. Machado (ed.), *Ethics in Management and Business*, Management and Industrial Engineering, https://doi.org/10.1007/978-3-031-82054-0_1

Adorno's claims about our social world are resolutely scathing and unforgiving: our social world, including the sphere of business, is a 'damaged', 'false', even 'radically evil' one (Adorno 2005a; Reeves and Sinnicks 2021; Nevasto 2021). Our social world is, he claims, so distorted and unfree that it is impossible not only to live well in it, but even to know what living a good or even a right life would be. Such claims are discomfiting and run against the grain of the moral-philosophical—and business ethics—traditions, which assume individuals can, at least in principle, live well or at least rightly in our social world. But that these claims are troubling does not entail that they are false; wishful thinking is the enemy of insight.

In this chapter we outline Adorno's critical ethical thought, troubling as it may be, and show how it bears on topics in business ethics, and on the more general orientation of the business ethics tradition. We then turn to a significant controversy in the interpretation of Adorno's ethics amongst scholars who advocate its importance for business ethics, and finally by sketching its implications for business ethics questions such as the possibility of good work. Taken together, these discussions reveal the unignorable, even if uncomfortable, relevance of Adorno's thought for business ethics.

2 Adorno's Critical Theory and Business Ethics

The central theme of Adorno's philosophy is a critique of what he perceived as the moral poverty of the modern social world. He argued that this world is inherently 'damaged' (Adorno 2005a), making it a hostile environment for human beings, stifling both their flourishing and their freedom. Adorno described the organizing principle of modern society as the "universal exchange relationship," where everything exists only in relation to something else (Adorno 1993, p. 26). This exchange principle stems from a basic drive within subjects to dominate objects (Adorno 2008, p. 9), which motivates categorizing people and things into abstract categories, increases the subject's power to classify, predict, and control.

This drive is crystalized in commodity exchange, reducing all things, including people, to exchangeable units. As a result, human beings and their needs become "mere things" subject to the laws of the marketplace (Zuidervaart 2007, p. 76). In this 'exchange society', everything and everyone is reduced to a means for something else, creating a world where people are defined by their economic roles as buyers and sellers. These roles reduce individuals to mere quantities. Consumers are seen as a quantity of purchasing power, and workers as a quantity of labouring power. To meet their basic needs, individuals are compelled to commodify their labour: an exchangeable unit determined by what is profitable, which often bears no relation to actual human needs.

If we treat people as exchangeable units, we risk 'reifying' them—seeing them as mere objects to be manipulated and controlled, rather than recognizing one another as unique individuals (see Jutten 2010). This reification contributes to the sense that modern society is 'unfree' (Adorno 2006, pp. 202–203) because it undermines the possibility of autonomy. That is, it prevents individuals from exercising self-governing control over their lives by forcing them to engage in activities that are heteronomous. However, rather than causing a revolt against present conditions, people have by and large adapted so thoroughly to this world that they have sacrificed the development of the rational capacities and impulses needed for autonomous engagement with both their psychic and social realities. Our experience is for the most part conditioned by distorting patterns of thought, feeling and action. This distortion is draining, and leaves little room for autonomous, critical experience and practice.

In this light, it is unsurprising that Adorno's work sits uneasily alongside mainstream moral philosophy, of the sort that has played the biggest role in business ethics. Indeed, Adorno argues that moral philosophy is possible only as a "critique of moral philosophy, the critique of its options and awareness of its antinomies" (Adorno 2000, p. 167). This involves an expansion of what moral philosophy is thought to be, and in that sense what business ethics is thought to be as well.

The mainstream of Anglophone moral philosophy poses questions such as 'what ought I to do?' addressed to an individual, and the 'ought' stands for a specifically ethical dimension of the world. Theories of business ethics, as frameworks for enquiry into how to morally or ethically assess and guide the behaviour of companies and their personnel, typically follow this model. While the meaning of the 'ought' varies between deontology, consequentialism, pragmatism, and existentialism, they for Adorno guilty of a similar mistake.

Mainstream moral philosophy is—to draw on Horkheimer's distinction—an example of 'traditional' as opposed to 'Critical' theory (Horkheimer 1989; see also Parker 2003). In contrast to traditional theory, Critical Theory asks after the conditions of possibility for posing questions in this manner. Traditional theory has to presuppose things such as the moral-psychological capacities of individuals and the justification of conceptions of the good. Interrogation of these opens up the relationship between socio-historical conditions and philosophical claims, which demands the suspension of thinking of these domains as methodologically distinct.

The rationale for this position derives from an attack on the notion of philosophical invariants (whether playing out in epistemology, ontology, or metaphysics) as they are thought to forge an identity between the concept and its object: "Dialectics is the consistent sense of non-identity. It does not begin by taking a standpoint" (Adorno 1973, p. 5). Adorno maintains that invariant principles hypostasize existing social reality by bracketing off some particular aspect of it, and then claiming it has objective, universal validity for our thought and action. Following Hegel, Marx, and Nietzsche, Adorno holds that the social conditions characteristic of specific (typically privileged) social groups reappear in the formal structures of philosophical works. Critical Theory is thus, for Adorno, an essential contributor to the tasks of philosophy: "The need to lend a voice to suffering is a condition of all truth" (1973, pp. 15–16).

This leads to a thoroughgoing 'metacritique' of rationalistic theories of morality. The thought is that moral knowledge presupposes virtue, which is itself not a product of reason, but of acculturation as a shared achievement. A moral agent has the capacity to acquire it, but moral capacity cannot emerge unless one lives among other such agents within institutions such as families and workplaces might foster it. This, Adorno argues, is not possible in mass capitalist societies. Adorno's argument to that effect is as follows: As moral knowledge as virtue takes the form of non-reflective situational awareness as to how to act, it presupposes knowledge of one's situation and its relative stability as a form of life. These presuppositions are unavailable to modern actors. Self-preservation enforces the calculative and strategic orientation in any area markets and economic-administrative systems penetrate. Even if impulses such as empathy are preserved, the lack of control individuals and organisations have of the social conditions of moral action typically render it void beyond narrow domain of private existence.

Nonetheless, we propose that this is not the whole story about virtue ethics with respect to its philosophical form. Recent work has brought together Neo-Aristotelian ethical naturalism (Foot 2001; Hursthouse 1999) and Adorno's Critical Theory (see Freyenhagen 2013; Reeves 2016; Whyman 2017). The crux of this debate is thought that if the human good inheres in our natures—as Neo-Aristotelians hold—it can secure an appeal that cuts across its current historically distorted form under capitalist societies, and this can combat both theoretical and practical nihilism and possibly buttress radical political claims.

While the good life is not practically available in current circumstances, this strategy makes use of those elements in Adorno's writings which emphasise the particular instances of moral badness and takes these as a guide to ethical normativity. Critical reflection on ordinary or common-sense experiences can itself be ethically salient, despite not yielding results to anything like the narrowly conceived task of modern moral theory. There is an ethically salient cognitive advantage in knowing badness, especially the worst kinds of badness (Peters 2015). Pursuing this line of thought allows us to register numerous ways in which Adorno's social critique identified, anticipated and diagnosed ethically bad features and developmental tendencies of the contemporary world of business, and of the discourse of business ethics that attempts to rationally redescribe it.

3 Adorno's Critique and Topics in Business Ethics

We now indicate what this changed, critical conception of thinking about business ethics might concretely amount to regarding key topics of business ethics.

3.1 Positivistic Management

Adorno might not be a 'moral philosopher' in the sense that aligns with traditional theory, but his critical comments on social phenomena are far from being ethically neutral. Indeed, given that his Critical Theory is deeply and intentionally an examination of the ethically salient societal structures around us, Adorno emerges as a valuable resource for critically analysing the ethical dimensions of organizational and commercial practices in capitalist systems. There is, after all, much in the world of business and management that is worthy of negativist critique.

Part of the reason for this is that a considerable portion of management research adopts a scientistic and technicist view of management. Management and organisation studies often fall prey to 'physics envy' (Thomas and Wilson 2011; Joullié and Spillane 2022). In practice, and given that management of people is central to the discipline, this means that it often becomes primarily a discipline concerned with the classification, evaluation, and control of individuals. Freeman and Newkirk acknowledge that "[i]mplicit in much of the management discussion is a mechanical, deterministic, positivistic view of business—a financial engine controlled by the machinery of scientific management" (2009, p. 117).

Adorno's critique of positivism is relevant here. Positivism takes knowledge produced by the natural sciences as the default form of human knowledge. It rests on two main assumptions: first, that the fundamental aspects of reality consist of states, events, and the constant conjunctions or universal regularities between them; and second, that knowledge is derived from observing these states and events and systematically organizing these observations through hierarchical if–then generalizations—laws—that allow deductions from these observations (e.g., Hempel and Oppenheim 1948; Popper 2002).

Adorno argues that positivism blurs the line between what is and what ought to be (Cook 2001). On such a view, positivism implicitly endorses the status quo and aims to gather predictive knowledge about human behaviour for the purpose of manipulation and control, much like the natural sciences use knowledge of law-like regularities to exert control over—indeed, even 'dominate' (Horkheimer and Adorno 2002; see also Cook 2011)—the natural world. Within the social realm, by reducing the world to empirical regularities, positivism overlooks the importance of unrealized potential, thereby stifling the possibility of meaningful change (Adorno 1973). Positivistic thinking thus encourages the belief that the social and psychological realities of the modern world are both inevitable and unchangeable, leading to their acceptance rather than critical examination.

This mindset tends to produce a dehistoricized, reductionist, and reified consciousness regarding human phenomena. Consequently, research influenced by positivism is likely to be deeply suspect on account of both implicitly validating a flawed social order while obscuring or deflecting attention from its systemic deficiencies (see Reeves and Sinnicks 2021).

3.2 *Consumer Culture*

A central aspect of Adorno's critique of consumer culture is the idea that people often embrace the trivialities offered by the 'culture industry' (Horkheimer and Adorno 2002), not because these align with their true needs, but because these distractions help them cope in a world that frustrates their genuine needs. This creates a paradox where these trivialities become real but ultimately false needs:

> People actually do not know what to do with their free time [because] they have been deprived beforehand of what would make the state of freedom pleasant to them. That state of freedom has been refused them and disparaged for so long that they no longer even like it. People need superficial distraction... in order to summon up the energy for the work that is demanded of them from the organization of society (Adorno 2005b, p. 172).

Moreover, this socially imposed triviality brings the qualities of labour, as enforced by a society governed by the exchange principle, into leisure itself (Adorno 2005b, p. 171). As a result, although free time is meant to contrast with work time ultimately comes to mirror it, with a similar structure imposed on different content: "surreptitiously the contraband of behavioural mores from work, which never lets go of people, is being smuggled in" (Adorno 2005b, p. 169).

The leisure and culture industries are designed to offer endless opportunities for superficial distraction. This was a central point in Adorno's critique of hobbies and the culture industry (2005b), but it has arguably become more pertinent in the present day. The advent of online streaming platforms is a case in point. It offers an almost endless stream of entertainment with addictiveness built in (for a fuller account, see Reeves and Sinnicks 2024).

The other side of this enforced triviality is what Adorno terms 'pseudo-activity,' superficially 'productive' hobbies that tend to be merely the negative image of the actually and meaningfully productive activities that the advanced division of labour withholds from most of us: "what they create has something superfluous about it… which in turn spoils any pleasure it might give" (Adorno 2005a, p. 172). While trivial distractions are a response to boredom that manifests a renunciation of imagination, pseudo-activities are a misplaced attempt to claw back a sense of genuine agency: "pseudo-activity is misguided spontaneity" (Adorno 2005a, p. 173). This pseudo-activity represents a desire for authentic action—marked by relative freedom from the external forces of capitalist society—that approximates genuine autonomy, yet it is easily subsumed by the logic of work (Reeves and Sinnicks 2024, pp. 293–294).

Schoneboom suggests that "paid work continually extends its reach and… leisure is caught up in the dynamics of intensification" (2018, p. 360). This is evident in our guilt when we fail to 'maximise the utility' of our free time, an experience which parallels the efficiency imperative typical of capitalism. The issue is not only that we should come to value 'idleness' (O'Connor 2018), but also that various undesirable aspects of work, particularly the fetishisation of conscientiousness regardless of the end to which it is put, can distort our understanding of what is truly worth pursuing, leading to the failure of the impulse for spontaneous activity.

3.3 Social Media and the Public Sphere

Another crucial way in which business shapes contemporary life is the influence of social media on the public sphere. Both news and participation in public conversation are increasingly mediated by a few dominant social media platforms with global reach. This reality has contributed to pathologies of democratic life such as 'fake news' and 'post-truth' culture, driving political polarisation, conspiracy theories and extremism of various stripes (Prakasam and Huxtable-Thomas 2021). In this light, the hopeful vision of the internet as a neutral facilitator of public debate is no longer credible.

Adorno would have found this reality unsurprising. Indeed, his description of the mechanism which govern public discourse offer a prescient account of the tendencies which have developed further in subsequent decades. As far back as the 1940s, Adorno identified the underlying irrationality and destructiveness present in the individual tendencies that 'the administered society' fosters (Adorno et al. 2019). In the 1960s, Adorno diagnosed right-wing extremism's tendency to understand itself as a defence of 'true democracy' (Adorno 2020, p. 24). Such tendencies are a conspicuous feature of today's social reality.

The subordination of news media to the logic of the culture industry had already eroded our capacities for rational discourse. News and public discourse are governed by the exchange principle. As a consequence, public discourse ends up being governed not by truth or rationality but by the untutored impulses at play in consumer preference. The result is a subjectivised and instrumentalised conception of public deliberation—an 'opinion delusion society' (see Adorno 2005b, pp. 105–122). In such a society, "the tendency becomes to view the average of subjective opinions—the average of subjective reason, as it were—as the pinnacle of objectivity" (Adorno 2019a, p. 57).

The upshot is a system promoting subjectivization, which glamorises arbitrary opinion. The role algorithms play in how news content is 'consumed' via social media is a telling example. In contrast to the editorial processes which shaped traditional news media, the algorithms of social media newsfeeds and recommendations operate out of view and beyond public scrutiny. The endless stream of content that we experience seems to emerge as if by natural growth, when in reality it reflects data profiles often procured without informed consent.

As a result of shaping ostensibly factual news according to partisan preferences, social media 'news feeds' invite the assumption that subjective preference is the measure of reality. This, in turn, underpins the rationalisation of a psychic defensiveness that fosters persecutory and destructive tendencies. Adorno suggests that the allure of subjectivist, 'opinion-delusion' thinking is the escape it provides from the responsibility to think for oneself. If the standard is mere subjective preference, error is ruled out and we are free to fit things into whatever narratives we find gratifying.

For Adorno, "the affective power inside people which resists [the idea of] objective, anonymous laws governing events over their heads is so immense that people will fall for this mechanism of personalization, even against their better judgment" (2019a, p. 35). Yet at the same time, Adorno rejects the idea of technology as a reified force, which suppresses the element of human agency in the development of technical means, which have after all been developed under specific social conditions and imperatives. Neither focusing on individuals nor on technology gets the structural totality that is at work properly into view.

4 Adorno's Critique of the Business Ethics Tradition

Adorno's Critical Theory bears not only on topics in business ethics, but on the broader philosophical orientation of the modern business ethics tradition. In this section we consider the Adornian critique of that tradition.

4.1 Critique of Moral Philosophy

As in contemporary moral and political philosophy generally, in the business ethics tradition Kantian ethical thought has loomed large. In addition to the explicitly Kantian work of the likes of Bowie (2017), Smith and Dubbink (2011), et al. the Kantian tradition is also embodied in work in Rawlsian (Norman 2015), contractualist (Luetge et al. 2016), and a range of other broadly deontological perspectives, including stakeholder theory (Gibson 2000). According to Kant, the basic epistemological story about knowledge is that cognition confronts our desires, but takes them to be formless and to follow an event-causal order. Our minds construct experience out of them, assuming a set of synthetic a priori categories. Kant has to appeal to a very curious notion to forge a link between the sensible material and cognition, namely "transcendental apperception", a process of which by his admission the subject remains unaware (Kant 2007, p. 140). A crucial consequence of this is that the moral will is, in the final analysis, understood to be unconditionally free to assign order of importance to our interest. Adorno argues that this makes it, on the one hand, mysterious how reason alone could motivate one to act, while our desires as mere raw material could not secure a stable relation between minds and worlds, as cognition of objects becomes indistinguishable from the mere imposition of reason's own order (Adorno 1973, pp. 221–223).

The metaphysical conundrum is echoed in Kant's moral philosophy. As Kant declares in the first formulation of the Categorical Imperative, action from moral duty is performed not only as acceptable to any rationally thinking human being, but as willing the principle followed to become a universal law: "I ought never to proceed except in such a way *that I could also will that my maxim should become universal law*" (Kant 2012, p. 17). In a Kantian spirit, principle 9 of The UN Global Compact for Corporate Sustainability states that "Businesses should encourage the development and diffusion of environmentally friendly technologies" (UN n.d.). An attempt to treat the 'should' in a strict Kantian fashion yields curious results. Innovations typically involve capital-intensive research and development, which cannot be initiated without the expectation of returns. But these cannot be reaped without future competitive advantage, which consists principally of exclusive knowing or having. This means that the owners and managers of developer companies could not possibly be motivated to act out of willing the new environmentally-friendly technologies to become universally diffused as it would undermine their *sine qua non* profitability expectations. That would be self-defeating towards the very essence of private enterprise and thus contrary to the categorical imperative. In other words, it would not only be unrealistic to expect business managers act as the Global Compact states, but, according to the categorical imperative, also morally impermissible.

The Kantian business manager faces a dilemma of either acting out imperatives other than morality, or not acting at all. It is similarly implausible to expect capitalist businesses to be run as "moral communities", where "the interest of every member of the community are equal to the interest of every other member" (Bowie 1998, p. 46). So far, this may sound familiar to those who have reservations about deontological ethics. However, the standard assessment of deontological ethics as "more ideal than a reality" with respect to business actors misses the crucial point (Crane et al. 2019, p. 103).

On the one hand, it is true that Kantian moral acts presuppose a world in which it is possible to unite subjective and objective interests, whereas capitalist business thrives on pitting subjects against one another, and all organisations against the objective expectations of the market as such. However, on the other hand, Kant is inadvertently perspicacious: it is antagonistic *reality*, in which one is formally free and freely giving oneself to be determined by forces outside one's control, which is articulated in painstaking detail in Kant's writings (Adorno 1973, pp. 214–221; 2000, pp. 151–152). Furthermore, Kant's admission that the ultimate connection-point between reason and impulse, located in a transcendental capacity all human beings possess, is an indirect way of admitting that it is a supra-individual process (Horkheimer 1993). What appears as metaphysically mysterious in Kant, doubles up as reified and mysterious in practices of social reality.

We propose in an Adornian spirit, that it is only on the basis of containing an essential truth about capitalism that Kant's moral thinking has been so highly productive for capitalist legal and market systems, and downstream as a business ethics mainstay. The Kantian 'ought' neatly transposes the relevant questions about the real conditions of possibility (ends of production, ownership and social value of business) of moral action onto conscience and hope. It is unique in allowing a simultaneous speculative moral utopianism and practical scepticism.

In other words, Kant's 'failure' to be useful to market agents is itself illuminating about market society. Business ethics scholars might benefit from Adorno's dialectical mode of reasoning by both recognising the incoherency and ideological function of Kant, but nonetheless persist with his thought, reading him against the grain, and ask: what is it about our economic system that simultaneously throws up the demand to think universally beyond particular individual or communitarian interests, and yet straddles agents in positions in which they cannot meaningfully act? If Kant's thinking is more ideal than one fitting reality, then perhaps it is reality that is at fault. What, if anything, could be done to make the individual interests cohere with the interests of all and vice-versa?

4.2 Critique of Political Economy

To answer these questions in principle requires a cold hard look at what is going on in our economies. Resources for that are diverse, but we offer "explanatory hypotheses" of sorts, via the intimate mentor to Critical Theory that is Marx's critique of political economy.

The objective social interest may well be posited to be happiness, efficiency, individual freedom or whatever else, but in our practical social life the reproduction of capital mediates them. In a sense, capital is the objective social interest within the world of its own making. What, one may ask, is wrong with that? Given this is the case, would it not be in in the objective interest to have more economic growth?

The difficulty, according to Marx, is that capitalism allows for the mediation of the subjective and objective only in a highly abstract form: as value, expressed in prices, as "a relationship between the individual labor of producers and the total labor of society" (Heinrich 2004, p. 55). It is only in market exchange where the subjective interest in some way connects with an objective interest. However, the objective interest is an enigma: the aims and organisation of the total labour of society do not correspond to any co-ordination, but rather follow pressures corresponding to expected returns on capital. In social terms capitalism is a type of self-contradiction as it can only preserve itself by intensifying the often-destructive competition between various units of capital, its agents, and workers.

Whilst such creative destruction can be the short-term interest of specific groups and individuals, the dynamic makes it impossible to achieve a conscious and deliberative approach to objective social goods, and thus it is often thought that optimally satisfying subjective preferences can take its place as the purpose of the economy (see, for instance, Heath 2014).

Such an assumption—inherited from economics textbooks—glosses over complicated issues. Adorno writes, "[b]ut a thought based on a need becomes confused if our conception of the need is purely subjective. Needs are conglomerates of truth and falsehood; what is true is the thought that wants the right thing" (1973, p. 93). On the one hand, there are things which human beings need, and things they do not but that may deceptively appear to be needed. The addict, for example, who is at some level convinced they need the drug or the gambling or whatever, is deceived by a distorted development of appetite, as may be clear to those around them, and even to them. But there are also needs that become actual because of the context yet are still false needs.

In one respect, what individuals need will be relative to the social world in which they live—for example, as we saw: managers may need to feel in control of their employees, as in the past supervisors felt in control of the machine; workers may need mindless and predictable distractions in their free time in order to recuperate from draining work and in order to stave off the bleak sense that their imaginative capacities are atrophying, etc. Often, when our needs go unmet, we adapt by channelling our needs into substitutes that can be met. We may actually need something in order to cope with a bad situation, and yet that need (and its fulfilment) may in turn be sustaining the bad situation that gives rise to it. The tendency to adaptation to a privative situation fuels the problem that our actual needs may often prop up the bad social situation that generates them.

The difficulty of separating the true needs from the actual but false needs is notorious, because of this tendency to adaptation, and because the articulation and satisfaction of needs takes place in alienated social relations, and it is these relations that we simultaneously reproduce whilst satisfying our actual, individual needs. The production of environmentally sustainable technologies is a true need in the sense that the planet must be sustained for life itself to go on. Nonetheless, it is arguable that lesser production as such could achieve similar or better results, thus making, say, electric vehicles potentially simultaneously a 'true' and a 'false' need: if they help to preserve the planet (and they may well do so), they do so only in and through preserving our capitalist social relations of production. The difficulty is that we cannot tell what the true objects of our needs are without a reflective insight into the aims of economic production as a whole.

4.3 Critical Business Ethics as Metacritique

Our earlier remarks on positivistic management, leisure consumption and social media exhibit this pattern of critique. Whereas the common sense of business ethics literature is that critique must be practical and propose improvements, Adornian critique rejects this demand in favour of a more curious and resilient commitment to the question of what it is live to well, which it poses to the concrete reality of our social world. This runs counter to the dominant demand in contemporary business ethics to make constructive recommendations. As Adorno puts it,

> One continually finds the word *critique*, if it is tolerated at all, accompanied by the word *constructive*. The insinuation is that only someone can practice critique who can propose something better than what is being criticized. By making the positive a condition for it, critique is tamed from the very beginning and loses its vehemence. (Adorno 2005b, p. 287)

Adorno's point is that the demand—ubiquitous in the business ethics tradition, and in the editorial demands of the top journals in the area—to offer some positive, constructive side threatens the integrity of critique. For we cannot assume a priori that it is possible to offer constructive recommendations, and even if possible, it is not a necessary condition of a valid critique.

Of course, in some circumstances "critique can proceed by way of confronting realities with the norms to which those realities appeal," because "following the norms would already be better" (Adorno 2005b, p. 287). But this will not always be enough, not least because it will not always be the case that following the norms will be better. If the appealed-to norms themselves are corrupt, following them may be worse than the norm-breaching reality, and there are contexts in which we are glad that the reality does not live up to the bad professed norms. Furthermore, realities typically appeal to numerous norms that may be in conflict not only with the reality but with each other. Hence, simply siding with the norms to which a reality appeals will not solve the problem. Critique cannot be insulated from the possibility of being led beyond such a comfort-zone.

The deeper form of critique will confront the reality with the fact that it appeals to values that it systematically, rather than accidentally or occasionally, flouts. This does not endorse those values any more than it endorses the reality that espouses them, but it challenges the whole—the reality to which values that it essentially flouts are essential. The conclusion of such a critique might be that the social whole is incoherent or contradictory insofar as it necessarily appeals to values it necessarily flouts. We would be interested then in why societies appeal to values they fail to live up to, and to decipher important aspirations in those values as well as what blocks their realisation.

This procedure is what Adorno calls 'metacritique'. Critical business ethics thus would involve a process of metacritical moral learning as unlearning. This might take in considering the extent to which ethical notions register social experiences but in distorted form, and the extent to which the dominant casting of certain problems as "ethical" is itself a diversion from tackling other questions—of participation, ownership, aims of production, etc. In parallel to what has been suggested as the meaning of the Frankfurt School for critical management studies, critical business ethics could proceed in the spirit of "embracing its constraining elements and using the contradictions it produces to move towards something new" (Cluley and Parker 2023, p. 1711).

But the question whether such an investigation itself presupposes some philosophical grounds is far from uncontroversial. Indeed, in the emerging literature on Adorno and business ethics there is a dispute about precisely how to interpret and extend Adorno's moral philosophy (in which the present authors have been interlocutors), to which we now turn.

5 Inverted Aristotelian or Resolutely Metacritical?

For Adorno, then, the idea that ethical questions can coherently be posed as issues of abstract principle is an expression of the functional needs of a socio-economic system of domination and reification, and their tasks. But given that Adorno rejects the standard ways of approaching ethics, and so business ethics, might there still be a way of providing an ethical basis for Adorno's Critical Theory in the language of moral philosophy? Most plausibly, Adorno's thought sits in the Aristotelian ethical naturalist tradition, although in a negative or inverted form (see Freyenhagen 2013; Reeves 2016).

Rather than try to offer discursive justification for particular normative claims, Aristotelian ethics tries to account for the possibility and reproduction of virtuous character. Recent Aristotelian accounts do this by trying to elucidate the natural normativity of the human sphere (e.g. Foot 2001; Hursthouse 1999), attempting to spell out the naturalistic status of the virtues as elements of the proper functioning of the human being given our specifically human, rational nature. On this view, ethical *qua* natural normativity bottoms-out in our (distinctively rational) nature, and the task of ethics is to clarify that.

It makes sense to interpret Adorno in this tradition for several reasons. First, naturalistic normative language is ubiquitous in his works: our social world and its inhabitants are "damaged", "deformed", "sick", "ailing", "mutilated", etc. (Adorno 2005a). Second, Adorno repeatedly rejects the idea that ethical claims can or should be discursively justified by appeal to more general principles. He writes,

> Nobody should be tortured; there should be no concentration camps… [These] lines are true as an impulse, as a reaction to the news that torture is going on somewhere. They must not be rationalised; as an abstract principle they would fall promptly into the bad infinities of derivation and validity. (1973, p. 285)

The implication is that ethical insights depend on the right sort of impulses or immediate reactions and cannot be justified by appeal to more abstract principles since those principles would anyway need discursive justification in turn, a chain of derivation that would anyway bottom-out in impulses sooner or later. Rather, "breaking off rationality at such places better serves reason than a kind of pseudo-rationality that erects systems where it is first and foremost a question of immediate reaction" (2005b, p. 304).

It is not that Adorno embraces an epistemology of immediacy or impulse, which would be just as arbitrary and abstract as the rationalism he rejects (cf. Hulatt 2014). Rather, while bodily impulses are the crux of ethical responses, these impulsive reactions are only ethically adequate insofar as they are properly tutored, and so are not unmediated by reason. So far, so Aristotelian.

However, while positive Aristotelianism assumes a social world that adequately actualises and nurtures our human nature, for Adorno, humanity as such is yet to be socially actualised, and so cannot be positively appealed to. For positive Aristotelians our communities can be assumed to be adequate to tutor our impulses, and thus allow us to develop the virtues, including within the context of business and the workplace (Audi 2012). For Adorno that is simply not the case. On the contrary, our social world tends to suppress, erode and distort our ethical impulses, such that genuine ethical reactions are difficult to cultivate. Hence, adequate reactions are the aberration rather than the norm: the "reconciliation of spirit, the union of reason and nature, as it survives in [the] impulse" (Adorno 2006, p. 237) depends on reflection so that the impulse has passed "through coercive identity and [been] mediated by it," (2006, p. 266) against the grain of the social world that depends on a distorting glorification of abstract principles.

Where positive Aristotelianism views our ethical impulses as redeemed by the adequacy of the community that explains them, Adorno's inverted Aristotelianism views our impulses as in the typical case distorted and undermined by the inadequate community that explains them. The redeemable ethical reactions are those negative impulses—experiences of resistance and rebellion to social ailing and suffering—that are not best explained as mere symptoms of the badness of our social world. It is our unactualized human potential that experiences and diagnoses of badness implicitly appeal to. And the whole account is redeemable insofar as Adorno's Critical Theory offers a more comprehensive and plausible explanation of our social world than rival accounts (Freyenhagen 2013).

However, there is disagreement about how this fits with Adorno's metacritical commitments. On one prominent view, the unactualized human potential that metacritique of abstract ethics and business ethics must implicitly invoke has to be more than a mere potential or "postulate" (Freyenhagen 2013, p. 243). For diagnoses of *ailing* to make sense, the unactualized potential must correspond to a real human nature or human life-form that has so far been suppressed or distorted, and metacritique must presuppose it even though we do not know what it would be like and so cannot appeal to it as a positive standard.

Moreover, it might be thought that from insights into the badness of our situation, extrapolations to the good, to the suppressed potential, the true human life-form, are—however fallibilistically—possible (see Reeves 2016). From the urgent need, say, to avoid a tornado, and the traumatic quality of such a need, quite a lot can surely be discerned about us—our vulnerability, our need for safety and a persisting secure base—and from this, various implications could be drawn (Reeves and Sinnicks 2022, 2023). Or, from the fact that massive economic inequalities tend to generate and reinforce social problems for which no obvious solution has been found, it might be extrapolated that the human life-form involves a need for self-determination, democratic co-determination, and so mutual respect and the social bases for that. Such extrapolations would of course be fallibilist, and the more concrete we try to make them the more fallible our extrapolations will be. But that is nothing strange—scientific knowledge is accrued through similarly fallibilist extrapolations and speculations and their testing. In this light, the idea that nothing about the good could be even fallibilistically extrapolated from insights into the bad seems an arbitrary and unjustified restriction on moral imagination.

As introduced earlier, the other reading emphasises the thoroughgoing metacritical character of Adorno's Critical Theory and his critique of foundationalism (for discussion, see Nevasto 2024). Adorno's metacritique of philosophy takes the framing of the question of ethics as a search for the ultimate good as misguided. Such questions are not resolvable by moral theories, debates among which are a mirror of society predicated on competition. While philosophers, like various stakeholders, will inevitably defend their particular conceptions of the good, it is not the task of critical theory to weigh-in on such debates. Rather, its aim is to raise to theoretical consciousness the ways in which individuals and communities are all reliant upon and mostly powerless against the dynamics of capital's accumulation and reproduction. Accordingly, philosophy befitting a broken world involves a paradoxical attitude: it must face up to the experience of its suffering squarely, with despair but without nihilism, meaning that it must come to terms with the necessity of holding on to a utopian impulse, whist conceding the impossibility of any concrete over-riding conception of the human good.

We next consider the implications of this debate for a particular topic in business ethics—the possibility of good work.

6 The Question of 'Good Work'

This interpretive dispute has implications for business ethics, including the question of good work. Is work something that we ought to wish away as an ideology of wrong life—should we want to be liberated *from* work? Or should we want the liberation *of* work from the distortion of our social world? And, if the latter, are some kinds of work closer to being good even within our distorted social world, or is that a hopeless humanist delusion bound to dilute critical aspirations?

For Adorno, modern work is by and large unfree and against our interests: it is mostly "time that is determined heteronomously" (Adorno 2005a, p. 167), and this shows up in various forms of suffering. One such form of suffering is the "feeling of powerlessness" (2005a, p. 171). Most people today lack the freedom to "seek out and arrange [their] work according to [their] own intentions" (2005a, p. 169). Even if there is a range of kinds of work made available to people by the market, this range does not necessarily include many desirable options, and anyway most people's actual options regarding the sorts of work they can do on the labour market are very restricted to options they do not themselves 'seek out' so much as reluctantly accept. But the fundamental problem is not the extent of subjective choice or control over one's work; it is, rather, that the objective qualities of available work which would be necessary to render such choice or control meaningful are largely absent:

> If one is born as a worker, unfreedom persists objectively despite the semblance of levelling and equalization... Every single one of us can, in spite of everything, also experience this when for example we find ourselves in a job-seeking situation. It will be experienced primarily in the fact that what is expected of us as someone who... has to sell themselves on the market is not what we ourselves would like; that is, we cannot actually realize our own possibilities and talent but must largely follow what is demanded of us. (Adorno 2019a, pp. 58–59)

Since the available forms of work are determined by the market, by what is and is not profitable, there is a disconnect between available work and fulfilling work in which we could realize our distinctive, personal, human qualities. For this reason, the modern work-world is predominantly a sphere of subjectively experienced powerlessness—particularly for those who are already most socio-economically vulnerable, deprived or marginalised, such as women (see Koskinen Sandberg et al. 2018) and ethnic minorities (see McDowell et al. 2014).

Another form of suffering characteristic of contemporary work is boredom (Fisher 1993; Johnsen 2016). Insofar as it is a symptom of privation rather than mistake, boredom must have to do with the intrinsic, objective quality of work activities: "boredom is the reflex reaction to objective dullness... boredom is objective desperation" (Adorno 2005a, p. 171) flowing from "the [existing] structure of labour... the technically rationalised character of which basically consists in the repetition of ideally identical processes and operations on the part of the labouring individuals" (2019b, p. 155). That is, workers' boredom registers the objectively tedious character of the work they are expected to perform today, and the impulsive protest against it.

On the inverted Aristotelian view, Adorno's diagnoses of the ills of modern work suggest that the corruption of work in our social world, rather than work as such, is the problem, and that good, undistorted work is in principle ethically valuable. Moreover, Adorno's critique of work is not merely negative: "we are merely encouraging the cause of untruth if we turn *prematurely* to the positive and fail to persevere in the negative" (Adorno and Mann 2006, p. 97, emphasis added), but that does not mean we should stop at the negative. From there, fallibilistic positive extrapolations present themselves.

That we suffer from the heteronomy of modern work plausibly implies that we have a need for autonomous activity. The feeling of powerless stands in stark contrast to the self-guided, virtuous execution of judgment, skill, and practical wisdom implied by autonomous work. And workers' struggles for more 'meaningful' or 'humanized' work seem an appropriate reaction against the objective dullness of work that boredom registers. These extrapolations point us, perhaps, toward a critical reconstruction of the 'craft conception' of work.

On this view, while most modern work is privative, work as such, irreducible to mere leisure, is of ineliminable ethical significance for us bearers of the human life-form. Thus, Adorno's critique takes up the intuition of the 'anti-work' tradition (Gorz 1985) while reframing it: work in wrong life is privative, but work as such is ethically significant. We should want the liberation *of* work from the distortions of wrong life.

On the resolutely metacritical view, however, more circumspect conclusions follow. For any prospect of theoretically establishing notions such as good, better or humanly fitting work are, on this view, illusory. After all, the notions "good", "genuine" or "craft" work are also values within contemporary culture that is committed to self-realisation as a basic human good, mediated by the logic of capital's reproduction. The notion of human *qua* human forms of activity have been co-opted into the mainstream of business education by humanist management theories for decades. As Adorno was keenly aware, searching for non-reified domains in the annals of philosophy is attractive precisely because reified social relations are disappointing. Nonetheless, that search remains committed to "a latent idealism, and a barrier to incisive critique of idealism" (1973, p. 93).

The concern is that the inverted Aristotelian view might place Adorno too close to the 'humanist' or 'idealist' positions he was so critical of. Given that accounts of meaningful or fitting human activity have been integrated into the mainstream of business education by humanist management theories for decades (see Pirson 2019 for an overview), it would appear there is no gap in the market for such accounts.

Moreover, we may wonder what justifies singling out some values as normative standards from which ethical and organizational practises can be legitimately judged? If we try to extrapolate ethical values from negative insights, do we not run the risk of fetishizing the abstract negation, the mere opposite of what is bad about our social world? For example, in *Minima Moralia* Adorno expresses scepticism about the nineteenth century ideal of the full actualisation and realisation of human powers, as the mere abstract negation of the stultifying and oppressive forms of work characteristic of industrial society, which secretly aligns with the societal principle of that society:

> The conception of unfettered activity, of uninterrupted procreation, of chubby insatiability, of freedom as frantic bustle, feeds on the bourgeois concept of nature that has always served solely to proclaim social violence as unchangeable, as a piece of healthy eternity... Perhaps the true society will grow tired development and, out of freedom, leave possibilities unused... (2005a, p. 156)

This presses home the point that we should be wary of extrapolating positive ethical values from current privations. On the resolutely metacritical interpretation, Adorno's critique of modern work forces us to ask important questions about the limits of ethical insight. Rather than exploring positive ethical implications, we are encouraged to reflect sceptically on the purpose and value of doing so in our wider socio-economic situation.

Moreover, the same logic of capital reproduction penetrates the very process of writing about such matters in collections such as the present one, as the career prospects of academics depend upon producing chapters such as these: "the critic himself is also measured only in terms of his marketable success" (Adorno 1982, p. 20). This presses the starkly metacritical question of how far attempts to extract positive extrapolations from Adorno's negative criticisms are shaped by the demands and expectations of the economics of present academic commentary, which is also part of the socio-economic world—and such questions ought to be taken seriously. Taking these issues seriously, we might give more weight to Adorno's occasional favourable remarks on "freedom from work" (e.g. Adorno and Horkheimer 2019, p. 16).

Against this, though, Adorno does often seem to take the ethical worth of work seriously, and to think that we can extrapolate ethical values from the current privations of work. For example, he insists that "[w]e must not succumb to the ideology of work, but we cannot deny that all happiness is twinned with work" (Adorno and Horkheimer 2019, p. 35), and, when his friend and mentor Max Horkheimer pushes the anti-work position—"The worst thing is to mix up work and happiness" (2019, p. 39)—Adorno explicitly refuses it, insisting that "work is also happiness, but one isn't allowed to say so" (2019, p. 39). Such passages suggest that we should want the liberation *of* work from the heteronomous and tedious corruptions of it in our social world, and hint that even in wrong life work can sometimes approximate to worthwhile, good work.

7 Concluding Remarks

While the dispute about the interpretation of Adorno in relation to business ethics remains an open one, we hope to have made clear that, on either view, Adorno's Critical Theory has rich and provocative implications for the question of good work, for several other central topics in business ethics, and for the practice of business ethicists more generally.

References

Adorno, T. W. (1973). *Negative dialectics*. New York: Continuum.

Adorno, T. (1982). *Prisms*. MIT Press.

Adorno, T. W. (1993). *Hegel: Three studies*. Polity.

Adorno, T. W. (2000). *Problems of moral philosophy*. Cambridge: Polity Press.

Adorno, T. W. (2005a). *Minima moralia: Reflections on a damaged life*. London: Verso Books.

Adorno, T. W. (2005b). *Critical Models: Interventions and Catchwords*. New York: Columbia University Press.

Adorno, T. W. (2006). *History and freedom: Lectures 1964–1965*. Cambridge: Polity.

Adorno, T. W. (2008). *Lectures on negative dialectics*. Cambridge: Polity.

Adorno, T.W. (2019a). *Philosophical Elements of a Theory of Society*. Cambridge: Polity.

Adorno, T.W. (2019b). *Ontology and Dialectics*. Cambridge: Polity.

Adorno, T. W. (2020). *Aspects of the New Right-Wing Extremism*. Cambridge: Polity.

Adorno, T. W. & Horkheimer, M. (2019). *Towards a New Manifesto*. London: Verso.

Adorno, T. W., & Mann, T. (2006). *Correspondence 1943–1955*. Polity.

Adorno, T. W., Frenkel-Brunswik, E., Sanford, R. N., & Levinson, D. J. (2019). *The Authoritarian Personality*. London: Verso Books.

Audi, R. (2012). Virtue ethics as a resource in business. *Business Ethics Quarterly*, 22(2), 273–291.

Beyes, T., & De Cock, C. (2017). Adorno's grey, Taussig's blue: Colour, organization and critical affect. *Organization*, 24(1), 59–78.

Bowie, N. E. (1998). A Kantian theory of capitalism. *Business Ethics Quarterly*, 8(S1), 37–60.

Bowie, N. E. (2017). *Business ethics: A Kantian perspective*. Cambridge University Press.

Cluley, R., & Parker, M. (2023). Critical theory in use: Organizing the Frankfurt School. *Human Relations*, 76(11), 1689–1713.

Cook, D. (2001). Adorno, ideology and ideology critique. *Philosophy and Social Criticism*, 27(1), 1–20.

Cook, D. (2011). *Adorno on nature*. Routledge.

Crane, A., Matten, D., Glozer, S., & Spence, L. (2019). *Business ethics: Managing corporate citizenship and sustainability in the age of globalization: 5th edition*. Oxford University Press.

Fisher, C. D. (1993). Boredom at work: A neglected concept. *Human relations*, 46(3), 395–417.

Foot, P. (2001). *Natural goodness*. Clarendon Press.

Freeman, R. E., & Newkirk, D. (2009). Business as a human enterprise. In Gregg, S. & Stoner, J. R. (Eds.), *Rethinking business management: Examining the foundations of business education*: 139–43. Princeton, NJ: ISI.

Freyenhagen, F. (2013). *Adorno's practical philosophy: Living less wrongly*. Cambridge: Cambridge University Press.

Gibson, K. (2000). The moral basis of stakeholder theory. *Journal of Business Ethics*, 26(3), 245–257.

Gorz A. (1985). *Paths to Paradise: On the Liberation from Work*. London: Pluto Press.

Granter, E. (2019). The Frankfurt School and critical theory. In Clegg, S. & Cunha, M. P. (Eds.) *Management, Organizations and Contemporary Social Theory* (pp. 223–243). Routledge.

Heath, J. (2014). *Morality, Competition, and the Firm*. Oxford University Press.

Hempel, C., & Oppenheim, P. (1948). Studies in the logic of explanation. *Philosophy of Science*, 12: 135–75.

Heinrich, M. (2004). *An introduction to the three volumes of Karl Marx's Capital*. NYU Press.

Horkheimer, M. (1989). The state of contemporary social philosophy and the tasks of an institute for social research. In S. Bronner and D. Kellner (Eds.), *Critical Theory and Society: A Reader*. New York: Routledge.

Horkheimer, M. (1993). *Between Philosophy and Social Science: Selected Early Writings*. MIT Press.

Horkheimer, M., & Adorno, T. W. (2002). *Dialectic of Enlightenment: Philosophical Fragments*. Stanford: Stanford University Press.

Hulatt, O. (2014). Normative Impulsivity: Adorno on Ethics and the Body. *International Journal of Philosophical Studies*, 22(5), 676–695.

Hursthouse, R. (1999). *On Virtue Ethics*. Oxford University Press.

Johnsen R. (2016). Boredom and Organization Studies, *Organization Studies*, 37(10), 1403–15.

Joullié, J. E., & Spillane, R. (2022). "Physics envy" in organisation studies: the case of James G. March. *Journal of Management History*, 28(2), 236–254.

Jutten, T. (2010). Adorno on Kant, freedom and determinism. *European Journal of Philosophy*, 20(4), 548–74.

Kant, I. (2007). *Critique of Pure Reason*. Trans. Kemp Smith. Palgrave MacMillan.

Kant, I. (2012) *Groundwork of the Metaphysics of Morals*. Gregor & Timmermann (eds.) Cambridge University Press.

Koskinen Sandberg, P., Törnroos, M., & Kohvakka, R. (2018). The institutionalised undervaluation of women's work: The case of local government sector collective agreements. *Work, Employment and Society*, 32(4), 707–725.

Luetge, C., Armbrüster, T., & Müller, J. (2016). Order ethics: Bridging the gap between contractarianism and business ethics. *Journal of Business Ethics*, 136(4), 687–697.

McDowell, L., Rootham, E., & Hardgrove, A. (2014). Precarious work, protest masculinity and communal regulation: South Asian young men in Luton, UK. *Work, Employment and Society*, 28(6), 847–864.

Nevasto, J. (2021). Adorno's critical moral philosophy and business ethics. *Business Ethics Journal Review*, 9(7), 40–46.

Nevasto, J. (2024). Critique and its foundations: on critical realism and the Frankfurt School. *Journal of Critical Realism*, 23(2), 169–186.

Norman, W. (2015). Rawls on markets and corporate governance. *Business Ethics Quarterly*, 25(1), 29–64.

O'Connor, B. (2018). *Idleness: A philosophical essay*. Princeton University Press.

Parker, M. (2003). Business, ethics and business ethics: Critical theory and negative dialectics. In M. Alvesson & H. Willmott (Eds.), *Studying management critically*: 197–219. Thousand Oaks, CA: Sage.

Pirson, M. (2019). A humanistic perspective for management theory: Protecting dignity and promoting well-being. *Journal of Business Ethics*, 159(1), 39–57.

Peters, M. (2015). 'The Zone of the Carcass and the Knacker'— On Adorno's Concern with the Suffering Body. *European Journal of Philosophy*, 23(4), 1238–1258.

Popper, K. (2002). *The logic of scientific discovery*. London: Routledge.

Prakasam, N., & Huxtable-Thomas, L. (2021). Reddit: Affordances as an enabler for shifting loyalties. *Information Systems Frontiers*, 23(3), 723–751.

Reeves, C. (2016). Beyond the postmetaphysical turn: Ethics and metaphysics in critical theory. *Journal of Critical Realism*, 15(3), 217–44.

Reeves, C. & Sinnicks, M. (2021). Business Ethics from the Standpoint of Redemption: Adorno on the Possibility of Good Work. *Business Ethics Quarterly*, 31(4), 500–523.

Reeves, C. & Sinnicks, M. (2022). Adorno's Critique of Work in Market Society. *Business Ethics Journal Review*, 10(1), 1–7.

Reeves, C. & Sinnicks, M. (2023). Needs, Creativity, and Care: Adorno and the Future of Work. *Organization*, 30(5), 851–872.

Reeves, C. & Sinnicks, M. (2024). Totally Administered Heteronomy: Adorno on Work, Leisure, and Politics in the Age of Digital Capitalism. *Journal of Business Ethics*, 193(2), 285–301.

Schoneboom, A. (2018). It makes you make the time: 'Obligatory' leisure, work intensification and allotment gardening. *Ethnography*, 19(3), 360–378.

Smith, J., & Dubbink, W. (2011). Understanding the role of moral principles in business ethics: A Kantian perspective. *Business Ethics Quarterly*, 21(2), 205–231.

Thomas, H., & Wilson, A. D. (2011). 'Physics envy', cognitive legitimacy or practical relevance: dilemmas in the evolution of management research in the UK. *British Journal of Management*, 22(3), 443–456.

UN (n.d). The Ten Principles of the UN Global Compact. Retrieved from: https://unglobalcompact.org/what-is-gc/mission/principles [accessed 06.11.24].
Whyman, T. (2017). Adorno's Aristotle critique and ethical naturalism. *European Journal of Philosophy*, 25(4), 1208–1227.
Zuidervaart, L. (2007). *Social philosophy after Adorno*. Cambridge: Cambridge University Press.

Confucian Diligence: A Core Ethical Value That Shapes Business Success in East Asia

Sophia Chia-Min Chou

Abstract To elucidate the economic success of the Four Asian Tigers in labor-intensive industries during the second half of the twentieth century, Redding (The spirit of Chinese capitalism, Walter de Gruyter, New York, 1993) characterized Confucian diligence as a high tolerance for repetitive work. However, this explanation falls short in accounting for the rapid technological advancements in East Asia in the early twenty-first century. Through a cultural-system analysis of Confucian thought, this chapter reveals that Confucian diligence encompasses not only *relentless effort* but also the dimensions of *ceaseless improvement* and *boundless passion*. Subsequently, this chapter examines how Confucian diligence manifests in contemporary high-tech businesses in East Asia and contributes to their organizational success. Ultimately, by redefining the construct of Confucian diligence, this chapter provides a cultural explanation for today's rapid technological development and economic success in East Asia.

Keywords Capitalism · Confucianism · Diligence · East Asian economic growth · Four Asian Tigers · Technology development

In the early twentieth century, Weber (1951) argued that Confucian values hinder the rise of capitalism in China. However, this theory faced challenges as East Asia's economies demonstrated unprecedented growth in the latter part of the century. The rapid economic expansion of the Four Asian Tigers—Hong Kong, Singapore, South Korea, and Taiwan—required a new cultural explanation. Scholars such as Redding (1993) suggested that Confucian ethics, particularly the value of diligence, played a crucial role in driving this remarkable economic development.

Redding (1993) characterized Confucian diligence as a high tolerance for repetitive and monotonous tasks, along with a strong commitment to long working hours, emphasizing its significance in the success of labor-intensive industries in this region.

S. C.-M. Chou (✉)
Graduate Institute of Management, National Taiwan Normal University, 31 Shida Rd., Da'an Dist., Taipei City 106, Taiwan
e-mail: sophia.chou@ntnu.edu.tw

C. Machado (ed.), *Ethics in Management and Business*, Management and Industrial Engineering, https://doi.org/10.1007/978-3-031-82054-0_2

This trait was notably observed in Hong Kong and Taiwan at that time. However, Japan, known for its advanced technologies and prominent brands such as Sony and Toyota, did not fit this portrayal, as high-tech success relies on innovation rather than repetitive work. Consequently, Redding excluded Japan from his analysis, arguing that despite sharing the same cultural origins, Japan and China have very different approaches to doing things. This view seems to be confirmed by historical developments, as China's economic ascent in the 1990s was primarily driven by the export of labor-intensive products, such as apparel, footwear, furniture, and toys (Mathai et al. 2016), aligning with Redding's depiction of Confucian diligence.

Yet, three decades after Redding's publication, the explanatory power of his characterization of Confucian diligence has waned in light of East Asia's technological advancements. Taiwan's TSMC and South Korea's Samsung have emerged as major players in advanced integrated circuit (IC) manufacturing, while China has become a significant force in advanced technologies such as telecommunications (Miller 2022). Evidently, during this period, South Korea, Taiwan, and China have increasingly mirrored Japan's technology-oriented model.

Put more directly, while Redding's (1993) characterization of Confucian diligence can well explain the labor efficiency of the Four East Asian Tigers in the late twentieth century, it was inadequate in accounting for Japan's technological power during that era. Today, this explanation not only fails to capture Japan's high-tech achievements in the 1980s but also overlooks the technological advancements in Taiwan, South Korea, and China over the past three decades. This research gap leads to two important implications. First, despite the differing timelines of their developments, these East Asian economies share common traits, growing from labor-intensive industries to technology-driven sectors, highlighting the need for a more integrated approach to understanding their economic growth. From a cultural perspective, Japan and the Four Asian Tigers belong to the broader East Asian cultural sphere, significantly shaped by Confucianism (Prescott 2015; Richey 2015). Rather than viewing Japan as fundamentally different from its regional counterparts, it is more accurate to recognize that Japan exhibited certain Confucian cultural characteristics earlier than the others. Second, since Redding's notion of Confucian diligence does not fully account for the shift from labor-intensive to high-tech industries in the region, it is crucial to revisit and expand our understanding of Confucian diligence to better reflect the current economic landscape in East Asia.

Hence, to address the gap in Redding's characterization of Confucian diligence, this chapter will reexamine the underlying dimensions of this construct. Furthermore, rather than excluding Japan from the analysis of East Asian economic development, this chapter will include it alongside other East Asian countries. Accordingly, this chapter will begin by reviewing the various scholarly perspectives on the impact of Confucianism on East Asian economic development over time. It will then explore the research methods used in cultural studies. Finally, the chapter will analyze the concept of diligence in Confucian teachings and examine its manifestations within the context of East Asia's economic progress. This approach ensures that this chapter's analysis is grounded in a robust conceptual framework, encompassing prior perspectives, research methodology, cultural legacy, and empirical cases.

1 Confucianism and East Asia's Economic Evolution

The interplay between cultural factors and a region's economic development has long been an intriguing topic (e.g., Redding 1993; Weber 1951, 2001). Sophisticated cultural systems are akin to a kaleidoscope, revealing various dazzling facets across different times. In the early twentieth century, facing the rapid rise of the West, Weber (1951) attributed East Asia's economic lag to Confucianism. However, in the late twentieth century, as East Asia experienced unprecedented economic growth through labor-intensive industries, Redding (1993) attributed this success, in part, to Confucianism. As this region achieves technological advancements in the early twenty-first century, Redding's characterization of Confucianism fails to adequately explain the region's transformation, highlighting the need for a more nuanced understanding of Confucian values. This section will explore how scholars have interpreted the economic development of this region over time through a cultural lens and why it is essential to reassess their analyses.

1.1 Max Weber's Perspective in the Early Twentieth Century

In the early twentieth century, Western nations were experiencing rapid capitalist development, while China notably lacked similar economic transformations. Max Weber sought to explain the rise of capitalism in the West and its absence in China.

In "The Protestant Ethic and the Spirit of Capitalism," Weber (2001) elaborated on how Protestant ethics, particularly diligence in worldly vocations, contribute to capitalist growth. He notes that Protestants may not initially recognize themselves as predestined "children of God," but they are able to discern their "election" through active engagement in worldly activities. Their "inner-worldly asceticism," a drive for secular success rooted in religious conviction, was a crucial element in the development of capitalism in the West.

Conversely, in "The Religion of China," Weber (1951) pointed out that Confucianism lacks this concept of "inner-worldly asceticism." Instead, Confucian ethics promote a rational acceptance of existing social realities rather than the transformative rational control seen in Puritanism. Therefore, Weber believed that it was Confucianism that inhibited the rise of capitalist thinking in China. Ultimately, Weber concluded that the distinct social attitudes shaped by these dominant religious traditions help explain why capitalism thrived in the West but failed to emerge independently in China.

Accordingly, Weber argued that while Confucianism and Puritanism appear superficially similar—both valuing wealth accumulation and self-discipline—they diverge in fundamental ways. He emphasized that Puritanism, with its focus on "inner-worldly asceticism," promotes rational control over worldly affairs and a rejection of irrationality, which ultimately paved the way for the rise of capitalism in the West.

In contrast, he contended that Confucianism, lacking this inner-worldly asceticism, did not create the conditions necessary for capitalism to emerge.

1.2 Gordon Redding's Perspective in the Late Twentieth Century

Unlike the economic stagnation of the early twentieth century, many East Asian economies, particularly the Four Asian Tigers—Hong Kong, Singapore, South Korea, and Taiwan—experienced remarkable growth in the latter half of the century. These economies attracted substantial foreign investment and capitalized on their low-cost labor force, accelerating unprecedented growth through labor-intensive industries. By adapting their strategies to meet changing global demands, they maintained an average annual growth rate of around 6 percent between 1960 and 1995 (Barro 1998).

The evolving economic landscape in East Asia prompted scholars to investigate the cultural factors contributing to the region's economic achievements. Interestingly, just as Max Weber turned to Confucianism in the early twentieth century to explain why capitalism did not independently emerge in China, the late twentieth century saw a renewed focus on Confucian values to explain the remarkable success of the Four Asian Tigers.

For instance, Redding (1993) highlighted the diligence inherent in Confucian work ethics. He noted that Chinese individuals have historically been perceived as hardworking, often producing more output than other ethnic groups. In addition to production efficiency, they tend to take on more challenging tasks and demonstrate a remarkable tolerance for repetitive and monotonous work. This work ethic is characterized by seriousness about tasks, a strong sense of responsibility, and perseverance. Redding attributed this exceptional work ethic to their obligations of filial piety and the demanding process of learning Chinese characters. In Confucian societies, individuals are expected to work diligently to support their families, which creates significant social pressure to exhibit high levels of commitment. Moreover, the need to memorize thousands of Chinese characters fosters perseverance among learners.

While Redding's characterization of Confucian diligence effectively explains the manufacturing and operational efficiency contributing to the economic success of the Four Asian Tigers in the 1990s, it falls short of elucidating Japan's emergence as a technological powerhouse in the 1980s. Companies like Sony and other Japanese tech giants showcased innovative technologies during this period (Miller 2022). Consequently, Redding asserted that although Japan shares a cultural heritage with China, its business approaches differ significantly. He referenced Joseph Needham's observation, stating, "Japan is as different from China as if it had been towed away to Europe and anchored off the Isle of Wight" (Redding 1993, p. 2). Therefore, he suggested that his description of Confucian diligence as a high tolerance to repetitive and monotonous work did not apply to Japan's tech boom in the 1980s.

1.3 East Asia's Technological Development in the Early Twenty-First Century

Thirty years after Gordon Redding's publication, Taiwan, South Korea, and China have undergone unprecedented advancements in technology and innovation, reshaping the East Asian landscape and calling for further cultural interpretation. For instance, Taiwan has emerged as a global leader in advanced semiconductor manufacturing, with companies like Taiwan Semiconductor Manufacturing Company (TSMC) at the forefront of cutting-edge technology (Miller 2022). Since 2018, TSMC has outpaced Intel in process technology (Chen and Huang 2021), playing a crucial role in supplying chips for a diverse array of devices, including smartphones, computers, and advanced AI applications (Miller 2022). Currently, TSMC produces over 90% of the world's most advanced semiconductors (Kristof 2024).

Similarly, South Korea has established itself as a powerhouse in consumer electronics and telecommunications, with industry giants such as Samsung and LG driving innovation in smartphones, televisions, and home appliances. Samsung has also become a leader in semiconductor manufacturing technologies, being one of the two companies that have surpassed Intel in this field (Miller 2022).

Likewise, China swiftly transitioned its dominant exports from labor-intensive products in the 1990s to electronics in the 2000s (Mathai et al. 2016). Today, its high-tech industries span telecommunications, artificial intelligence, electric vehicles, and battery technologies (Miller 2022).

This transformation marks a significant departure from the traditional East Asian economic model of the late twentieth century (with the exception of Japan), which prioritized labor-intensive industries. Instead, recent technological developments in these East Asian economies closely mirror Japan's achievements in high-tech industries, signaling a shift toward more complex and innovative economic activities.

While Redding's initial analysis of Confucian work ethics adequately explained the earlier success of these East Asian economies, it is now insufficient to account for the current technological landscape in this region. After all, mere perseverance in repetitive and monotonous tasks cannot drive such advancements. This suggests that there may be important aspects of Confucian culture that scholars have yet to fully explore or recognize.

1.4 The Roles of U.S. Policies and Confucian Culture

In addition to cultural factors, it is essential to recognize the significant impact of U.S. policies on East Asian economic development. Following World War II, the United States helped Japan establish a peacemaking constitution and transferred transistor technology, aiming to transform Japan from a military power into a democratic, capitalist nation (Miller 2022). By the 1980s, Japanese companies had surpassed their American counterparts in the semiconductor, automobile, and consumer electronics industries, offering lower costs and higher quality. This phenomenon suggests that, while Japan's economic development was influenced by U.S. policies, its outstanding achievements in the 1980s indicate potential cultural differences between Japan and the West.

Similarly, in an effort to prevent the spread of communism among its Asian Pacific allies, the U.S. government facilitated the transfer of assembly lines for electronic products and semiconductors to South Korea, Taiwan, and Southeast Asia during the latter half of the twentieth century (Miller 2022). By the late 2010s, Taiwan's TSMC and South Korea's Samsung surpassed Intel in process technology, producing the world's most advanced microprocessors. In 2020, TSMC was invited to establish fabs in the U.S., followed by Japan in 2021. However, while TSMC's expansion in Japan progressed smoothly, it encountered significant challenges in the U.S. Despite starting construction of its Japanese factories about a year later than in the U.S., by February 2024, TSMC had inaugurated its Japanese factory, whereas its U.S. fabrication site faced delays and remained under construction (Zhou 2024). This further underscores the complex cultural differences between East Asia and the West in technological development.

The evolving dynamics of East Asian economies encourage us to delve deeper into the cultural factors underlying this region's technological progress. By examining these phenomena, we can unravel the complexities of how Confucian values may intertwine with modern innovation, leading to a richer understanding of the cultural foundations that support contemporary economic success in the region. Nevertheless, it is crucial to first consider how to approach the study of these cultural phenomena. While research methods can enhance scientific rigor, they may also constrain scholars' perspectives. Thus, it is worth exploring why Weber (1951) and Redding (1993) did not foresee the technological advancements in East Asia in the early twenty-firstcentury.

2 Research Methods of Cultural Studies

While many cultural studies have focused primarily on observable phenomena, scholars have cautioned that comprehensive cultural analysis must address two levels: cultural systems and empirical facts (Archer 1995, 2005; Hwang 2015a, b). Sociologist Archer's (1995, 2005) concept of *analytical dualism* emphasizes the need to distinguish between two levels of analysis: *structure* and *agency*. In her framework, structure contains *cultural systems*, while agency encompasses *socio-cultural interactions*. Cultural systems encompass the shared beliefs, values, and symbolic meanings within a society, whereas socio-cultural interactions involve how individuals influence and are influenced by these systems. This dualistic perspective enables a deeper exploration of how pre-existing structures constrain or enable individuals, whose actions can either reproduce or transform these structures over time. Archer critiqued the conflation of structure and agency, arguing that it limits our understanding of social phenomena. By analyzing these levels separately, researchers can more effectively examine the internal causal dynamics that shape social processes, enhancing our comprehension of the complex interplay between cultural systems and human agency.

Cultural psychologist Hwang's (2015a, b) *cultural-system approach* aligns with Archer's *analytical dualism*, emphasizing the importance of not restricting cultural studies to merely observable phenomena. Hwang contended that neglecting deeper cultural systems can obscure our understanding of experiential realities. Conversely, understanding cultural systems as a foundation for investigating empirical phenomena offers crucial insights into empirical studies. For instance, the original principles of Confucius and Mencius differ significantly from the authoritarian interpretations imposed by later emperors, illustrating the distinction between cultural systems and socio-cultural interactions.

From this perspective, a deep understanding of cultural systems can illuminate the dynamics of economic development at the socio-cultural interaction level. In contrast, an analysis that focuses solely on observable phenomena may overlook crucial aspects of cultural systems, which can manifest differently across various contexts. Building on this approach, this section will explore how Max Weber analyzed the rise of capitalism in the West (Weber 2001) and the absence of its independent development in China (Weber 1951), as well as how Redding (1993) explained the economic boom of the Four Asian Tigers in the latter half of the twentieth century.

2.1 Max Weber's Account of the Rise of Capitalism in the West

In "The Protestant Ethic and the Spirit of Capitalism," Weber (2001) introduced a groundbreaking approach to understanding economic activities through a cultural lens. He challenged the prevailing notion that individuals are purely rational economic actors, highlighting the need to examine the cultural and moral frameworks underlying economic behavior. While Weber acknowledged that culture alone is not sufficient for economic development, he argued that it plays a crucial role in shaping economic dynamics. Specifically, he posited that capitalist growth relies on the emergence of an entrepreneurial class, which is supported by a moral charter rooted in religious principles (Redding 1993). This relationship underscores the importance of cultural elements within the "spirit" of capitalism, offering a more nuanced understanding of economic phenomena.

Weber's analysis of Protestantism, particularly Calvinism, shows how religious teachings shaped attitudes and behaviors conducive to capitalism. His dual focus on both doctrine and practice reveals that the Protestant values of hard work, frugality, and a sense of calling significantly contributed to capitalist development. By linking religious beliefs to economic behavior, Weber provided a comprehensive framework that illuminates the complex interplay between ideology and practice, demonstrating how cultural factors can shape economic systems.

Weber's methodological approach demonstrates an integration of both cultural systems and the empirical world. He did not merely focus on observable economic activities; rather, he delved into the underlying cultural systems, such as religious doctrines, that influenced business endeavors. By examining how Protestant values like frugality and diligence were embedded in cultural systems and manifested in real-world economic behaviors, Weber provided a dual-layered analysis that connects abstract cultural principles with concrete empirical phenomena. This approach highlights the importance of understanding both levels to fully grasp the complexities of economic development.

2.2 *Max Weber's Explanation for the Absence of Independent Capitalist Development in China*

Weber (1951) argued that Western Puritans exhibited a characteristic of "inner-worldly asceticism," which he believed was absent in Confucianism. He contended that this distinction contributed to the rise of capitalism in the West, whereas capitalism failed to emerge independently in the East. However, historian Yü (2021), in his book "The Religious Ethic and Merchant Spirit in Early Modern China," challenged Weber's view by arguing that Confucianism does, in fact, embody inner-worldly asceticism through its emphasis on diligence and self-cultivation, guided by a belief in "Heaven"—a set of cosmic moral principles governing the universe and human affairs. According to Yü, capitalism did not develop independently in China because ancient Chinese governance failed to establish "reliable laws," one of Weber's (1981) six criteria for capitalism as described in "General Economic History." In short, Yü argued that it was the lack of a rational political and legal system, rather than the absence of inner-worldly asceticism in Confucianism, that hindered the development of modern capitalism in China.

As the issue of "reliable laws," described by Yü, gradually diminished with the advancement of East Asian governments, the cultural traits of Confucian inner-worldly asceticism began to flourish. This shift contributed to the rise of the Four

Asian Tigers in the second half of the twentieth century, initially through the development of labor-intensive industries. As capital accumulated, merchants and engineers in the region applied these cultural characteristics to technological innovation, leading to the emergence of a high-tech economy in the early twenty-firstcentury.

It is important to note that while Max Weber characterized Confucianism based on the ordinary lives of Chinese people, Yü offered a more nuanced portrayal by analyzing both the Confucian cultural system and the business practices of Confucian merchants. His analysis spans from the cultural-system level to the socio-cultural interaction level, providing a deeper understanding of Confucianism's influence on economic development in East Asia. This demonstrates that when scholars attempt to explain current conditions solely at the phenomenological level, they risk overlooking cultural meanings that may not have fully manifested due to underdeveloped temporal and spatial contexts. Such an approach may cause cultural studies to merely follow the fluctuations of economic phenomena. Therefore, it is crucial not only to analyze the phenomena themselves but also to examine the doctrinal aspects of the cultural system that underpin them.

2.3 Gordon Redding's Explanation for the Rapid Economic Development of the Four Asian Tigers in the Late Twentieth Century

Similarly, Redding's (1993) analysis of Confucianism and its impact on East Asian economies takes a more immediate approach, equating Confucianism with the daily lives of Chinese people rather than deeply investigating the original Confucian cultural system. Redding (1993, p. 43) asserted, "To understand Confucianism, it is better to begin by examining its function in society and how it has provided a particularly Chinese form of order linking the individual and the state." By focusing on the societal function of Confucianism, Redding explored how Confucian values manifest in social structures, yet his analysis risked overlooking essential elements of the Confucian cultural system that may remain latent in certain historical and social contexts. More specifically, by concentrating primarily on concrete practices and social structures, Redding may miss nuanced cultural factors that influence behavior and societal development. These cultural elements can remain dormant until favorable conditions allow them to flourish. Therefore, a more holistic exploration that incorporates both doctrinal and practical dimensions could offer a deeper understanding of how Confucianism shapes economic behavior in East Asia.

The evolving interpretation of Confucianism's influence on East Asian economic development—from the early twentieth century to the present—illustrates the limitations of analyses focused solely on the socio-cultural interaction level. An analysis centered exclusively on economic phenomena in East Asia risks overlooking foundational elements of the Confucian cultural system, which can manifest differently across various contexts. It is crucial to first explore the cultural foundations established by ancient philosophers before examining how contemporary East Asians apply these principles of Confucian diligence in their work ethics. A deep understanding of Confucianism at the cultural-system level can illuminate the dynamics of East Asian economic development at the socio-cultural interaction level.

In summary, to avoid the limitations of cultural explanations that are overly tied to specific economic contexts, as seen in the works of Weber (1951) and Redding (1993), this chapter advocates for a dual approach that integrates both cultural systems and socio-cultural interactions, drawing on the work of Archer (1995, 2005) and that of Hwang (2015a, b). This approach enables researchers to examine Confucian diligence within the framework of original Confucian teachings, deepening our understanding of its relevance and application in contemporary society. The following section will first explore the cultural system underlying Confucian diligence before analyzing its manifestations in modern East Asia.

3 Diligence in Confucian Classics—A Cultural System Analysis

In analyzing the cultural system of Confucian diligence, it is essential to examine the teachings of Confucius and Mencius, as they form the bedrock of Confucian thought (Chan 1963). However, it is also important to recognize that Confucianism did not emerge in a vacuum; it was significantly influenced by the ancient tradition of sage-kings (Chou 2024a). Both Confucius and Mencius highly esteemed ancient sage-kings, including Yao, Shun, Yu, Tang, King Wen, King Wu, and the Duke of Zhou.

For instance, Confucius once said, "The rulership of Yao was so magnificent! He was so sublime that even though there is nothing as great as Heaven, he could accord with it. His greatness was so boundless it is beyond description. His efficacy was amazing, his writings were enlightening" (*Analects* 8:19; Muller 2021). Confucius also held the Duke of Zhou in high esteem for his remarkable achievements in establishing a sophisticated system of rituals and music that educated and harmonized the nation. In his later years, Confucius expressed a sense of loss for no longer being able to dream of the Duke of Zhou (*Analects* 7:5), underscoring the deep respect he held for this sage-ruler. Likewise, Mencius often made laudatory reference to Yao and Shun (*Mengzi* 3A:1).

These revered ancient sage-kings exemplified diligent governance. Their actions and virtues set a precedent that shaped the moral framework of Confucianism. Therefore, this section will examine the dimensions of Confucian diligence through the actions of ancient Chinese sage-kings and the teachings of Confucius and Mencius, aiming to provide a nuanced understanding of this construct. In short, an in-depth analysis of the Confucian cultural system reveals that the concept of Confucian diligence involves not only *relentless effort*, as highlighted by Redding, but also two additional dimensions: *ceaseless improvement* and *boundless passion*. These aspects will be elaborated as below.

3.1 Relentless Effort

Sage-kings in ancient China were extremely diligent in their governance, embodying the principle of relentless effort. A prime example is Yu the Great around the twenty-firstcentury BCE. Modern scientific research indicates that, beginning in the twenty-fourth century BCE, the Northern Hemisphere experienced a significant drop in temperatures, leading to climate abnormalities (Baillie and McAneney 2014). As the climate shifted toward warmer conditions, accelerated glacial melting resulted in widespread flooding in many river basins where ancient populations resided. Historical accounts of ancient China describe the floods as severe (*Shiji*—Annals of Xia). The floods lasted for many years, causing extensive land submergence and displacing large numbers of people. Yu was tasked by Emperor Shun to alleviate the catastrophic floods that ravaged ancient China

Yu's remarkable dedication to flood control is highlighted in historical records. He traveled extensively and worked on the problem for over thirteen years, even passing by his own home without entering. His relentless efforts ultimately led to the successful management of the floods, earning him the throne from Emperor Shun and founding the Xia dynasty. Yu's story serves as a powerful testament to the relentless effort valued in Confucianism, illustrating how unwavering commitment and perseverance are essential for achieving long-term goals.

Yu's Xia dynasty was eventually overthrown by the Shang dynasty, which was later defeated by the founders of the Zhou dynasty in the eleventh century BCE, including King Wen, King Wu, and the Duke of Zhou. These leaders also exemplified admirable diligence. For instance, to ensure the happiness and harmony of his subjects, King Wen was so devoted that he rarely took time to eat. He also avoided excessive pursuits, such as hunting (*Shujing*—Against Luxurious Ease).

Similarly, in his quest to overthrow the ruthless Shang ruler, King Wu (King Wen's son) demonstrated great courage by launching a war to alleviate the suffering of the populace, even in the face of ominous divination signs (*Shuo Yuan* 15:16). His perseverance in overcoming challenges ultimately led to the defeat of the merciless Shang king.

King Wu passed away three years after the establishment of the Zhou dynasty. The Duke of Zhou (King Wen's son and King Wu's younger brother) played a vital role in assisting governance during the early Zhou period. His diligent support of the young King Cheng (King Wu's son) and his efforts to consolidate the Zhou dynasty's power underscore his unwavering commitment to effective administration. A notable example of his dedication was his practice of welcoming talented individuals to his court. He was deeply committed to attracting and retaining capable officials, so whenever they arrived during his meals, he would immediately spit out his food, ensuring he was ready to receive them without delay (*Shiji*—House of Lu Zhougong).

This behavior exemplifies the lengths to which the Duke of Zhou went to serve his state and fulfill his responsibilities. Furthermore, the Duke of Zhou frequently advised King Cheng to avoid indulging in luxurious ease and instead to emulate King Wen's diligence (*Shujing*—Against Luxurious Ease).

Influenced by these ancient sage-kings, Confucius also emphasized the importance of relentless effort in his teachings. When his disciple Zi Lu asked about how to govern. Confucius said, "Lead the people and work hard for them." When Zi Lu continued to ask for further advice, Confucius said, "Be not weary in these things" (*Analects* 13:1; Muller 2021; Legge 1861b). Confucius also advocated the value of relentless effort in learning. He famously stated, "If someone else achieves it in one try, I will strive to achieve it a hundred times; if someone else accomplishes it in ten tries, I will aim for one thousand" (*Zhongyong* 20). Furthermore, he criticized his disciple Zai You for sleeping during the day, implying that a lack of dedication was unacceptable (*Analects* 5:10).

Similarly, Mencius emphasized the importance of relentless effort. He noted that even easy-to-grow crops cannot flourish if subjected to a single day of warmth followed by ten days of cold, illustrating that consistent and stable effort is essential for growth (*Mengzi* 6A:9). He further illustrated this point by stating that well-trodden mountain paths eventually become roads, but if they are neglected for a period of time, they will be overgrown with weeds, emphasizing the necessity of persistent effort (*Mengzi* 7B:21). Additionally, Mencius cautioned against the dangers of insufficient effort, declaring, "Working at a project is like digging a well. If you dig sixty feet and stop without hitting water, it is just like abandoning the whole well" (*Mengzi* 7A:29). This metaphor highlights the critical importance of continuous and consistent effort, as even a single lapse in persistence can jeopardize the entire endeavor.

Through the examples of ancient sage-kings and the teachings of Confucius and Mencius, it is clear that the concept of relentless effort is deeply embedded in Confucian thought, highlighting the significance of unwavering dedication and perseverance in both historical and philosophical contexts. Although this principle resonates with Redding's view of Confucian diligence, it is important to note that Confucian diligence encompasses more than just relentless effort.

3.2 Ceaseless Improvement

In addition to their relentless efforts, ancient sage-kings exemplified a profound commitment to ceaseless improvement in their practices. A prime example is Yu the Great, who learned valuable lessons from his father Gun's failure to control floods through barrier construction. Gun's approach of building dams to block floodwaters ultimately proved ineffective, as the sheer force of the floods frequently overwhelmed these barriers, resulting in catastrophic damage and suffering for the people. Rather than persisting with ineffective methods, Yu innovatively shifted his approach by focusing on dredging and diverting water, ultimately leading to his success in mitigating the floods (*Shiji*—Annals of Xia). His ability to adapt and refine methods in response to challenges highlights the value of continuous improvement.

Similarly, the founders of the Zhou dynasty demonstrated the spirit of ceaseless improvement. For instance, during his seven years of imprisonment by the ruthless Shang king, King Wen expanded the Book of Changes (*I-Ching*) from 8 trigrams to 64 hexagrams (*Shiji*—Annals of Zhou). He wrote interpretations for each hexagram, known as "*guaci*," to elucidate their meanings. Following King Wen, the Duke of Zhou further developed this work by adding "*yaoci*," or interpretations for each line within the hexagrams. Their significant contributions to the *I-Ching* reflect their unwavering commitment to continuous improvement and intellectual advancement.

Furthermore, King Wu and the Duke of Zhou illustrated ceaseless improvement by reflecting on the downfall of the Shang dynasty. Shang kings frequently relied on rituals and sacrifices to appease deities in hopes of securing blessings for their rule. However, the dynasty ultimately collapsed under the tyranny of its last monarch. In light of the Shang's downfall, King Wu and the Duke of Zhou reasoned that only benevolent rulers dedicated to the well-being of their people could receive Heaven's blessings and sustain effective governance (Pines 2023). This philosophical shift marked a significant advancement in political thought, emphasizing the ruler's responsibility to their subjects and ultimately contributing to the establishment of the longest dynasty in Chinese history, which lasted for approximately 800 years.

Many ancient Chinese texts underscore the value of ceaseless improvement. For instance, the *I-Ching* states at the very beginning, "Heaven moves vigorously; accordingly, the noble person constantly strives for self-improvement," highlighting the necessity of ongoing self-cultivation. Similarly, the Great Learning notes that an ancient sage-king, Tang of the Shang dynasty, had his bathing tub engraved with the words: "If you can one day renovate yourself, do so from day to day. Yea, let there be daily renovation" (*Da Xue* 6; Legge 1861a). These teachings reflect a deeply ingrained cultural value in ancient China that revered personal growth and ceaseless development.

Likewise, Confucius actively encouraged his students to engage in continuous learning and the ongoing cultivation of their moral character. For instance, his disciple Zi Gong believed it commendable for a poor man not to grovel and for a rich man not to be arrogant. However, Confucius pointed out that higher ideals still exist—specifically, a poor man can learn to find contentment, while a wealthy man can cultivate a love for propriety (*Analects* 1:15). Therefore, another of his students, Zengzi, committed to daily self-reflection to ensure the continuous development of his moral character (*Analects* 1:4). This emphasis on continuous improvement underscores that Confucian diligence encompasses not only relentless effort but also ceaseless improvement.

As previously mentioned, both Confucius and Mencius held the traditions of ancient sage-kings in high regard. They encouraged Confucian scholars to look up to the perfect virtues of these sage-kings as a means to cultivate their own moral character. For instance, Confucius' disciple Yan Yuan aspired to emulate the ancient sage-king Shun, renowned for his exemplary virtues (*Mengzi* 3A:1). However, achieving

Sagehood was no easy feat; even Confucius humbly acknowledged that he did not consider himself a sage (*Mengzi* 2A:2). This acknowledgment highlights the challenges inherent in striving for such lofty ideals. By viewing sage-kings as role models and aspiring toward goals that seemed almost unattainable, Confucian scholars recognized the necessity of putting in relentless effort for ceaseless improvement. In the pursuit of moral excellence and the welfare of the populace, they aspired to improve themselves daily and cultivated the perseverance and dedication required for their challenging mission. As Zengzi stated, such pursuit would persist relentlessly until the end of Confucian scholars' lives (*Analects* 8:7).

3.3 Boundless Passion

According to Mencius, ancient sage-kings were not only diligent in their governance but also driven by an unwavering passion for the well-being of their people. They genuinely rejoiced in the happiness of the populace and were deeply concerned about their sorrows, fostering a reciprocal relationship where the people responded in kind (*Mengzi* 1B:4). Furthermore, noble rulers found joy in effective governance that brought peace and happiness to their subjects (*Mengzi* 7A:21), illustrating that their leadership was not merely a duty but also a source of profound fulfillment. Thus, Mencius' insights reveal that the diligent efforts of sage-kings were fueled by a lasting commitment to the welfare of their people, demonstrating a harmonious blend of responsibility and passion in their rule.

Likewise, Confucius said that he enjoyed learning and teaching so much that he never studied without getting bored and taught without getting tired (*Mengzi* 2A:2). Confucius also encouraged his students to discover joy in their learning journeys, famously asking, "Is it not a joy to learn and practice what you have learned?" (*Analects* 1:1). He further asserted, "To know is not as good as to love; to love is not as good as to delight in" (*Analects* 6:20). This highlights his belief that genuine enthusiasm and joy in learning far surpass mere knowledge, inspiring individuals to engage deeply with their studies and pursuits.

Confucius also illustrated the transformative power of boundless passion through his own experiences. He expressed that his zeal for knowledge was so intense that he would forget to eat, lose track of his worries, and remain unaware of the passage of time and the approach of old age (*Analects* 7:18). This notion underscores how passion can become all-consuming, allowing individuals to immerse themselves completely in their endeavors and pursuits.

Similarly, Mencius emphasized the joy inherent in moral cultivation. He noted that a tireless joy in the virtues of benevolence, righteousness, self-consecration, and fidelity embodies the nobility of Heaven (*Mengzi* 6A:16). Furthermore, he stated, "Honor virtue and delight in righteousness, and so you may always be perfectly satisfied" (*Mengzi* 7A:9; Legge 1895). For Mencius, the pursuit of moral excellence is not a painful process; instead, it is a fulfilling journey that brings happiness and contentment. Furthermore, while ancient sage-kings like Yao and Shun may have appeared to be unattainable role models to many (Chou 2024a), Mencius encouraged people to "delight in the Way of Yao and Shun" (*Mengzi* 5A:7), suggesting that their virtues and principles could inspire individuals in their pursuit of moral excellence. Mencius' perspective on passionate dedication encourages individuals to view their efforts as a path to personal growth and a deeper connection with their values.

Therefore, in Confucian thought, boundless passion constitutes an integral aspect of Confucian diligence. It motivates individuals to engage in relentless effort and ceaseless improvement across different domains, including learning, moral cultivation, and governance.

In summary, this section employs Hwang's (2015a, b) cultural-system approach to demonstrate that the concept of Confucian diligence extends beyond mere relentless effort to encompass ceaseless improvement and boundless passion. Confucian diligence has been shaped by the examples of ancient sage-kings and Confucian scholars' aspiration toward such lofty role models. Ancient sage-kings embodied all three aspects of Confucian diligence: relentless effort, ceaseless improvement, and boundless passion. Confucian scholars, inspired by these venerable figures, passionately sought to emulate them—a challenging and often elusive goal that required relentless effort and continuous improvement. Thus, Confucian diligence encompasses not only sustained effort and continuous self-improvement but also a profound passion for achieving moral and societal excellence, as illustrated in Fig. 1.

Among the three dimensions of Confucian diligence, relentless effort alone is sufficient to explain the economic success of the Four Asian Tigers in the 1990s. However, it does not fully account for the innovations seen in Japan during the 1980s and the rapid technological advancements in Taiwan, South Korea, and China in the 2020s. As mentioned earlier, the manifestation of cultural systems is shaped by historical and social frameworks, which influence how values are perceived and

Fig. 1 The three dimensions of confucian diligence

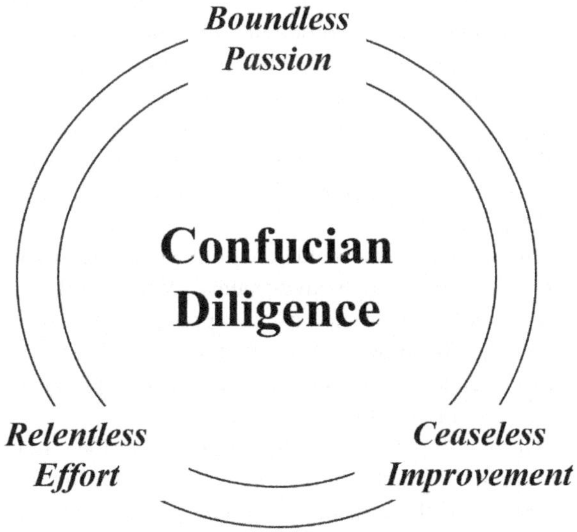

practiced across different societies and periods. Hence, the following sections will examine whether this multi-dimensional interpretation of Confucian diligence within the Confucian cultural system applies to two high-tech cases in East Asia.

More specifically, this chapter will analyze two high-tech cases to elucidate how Confucian diligence manifests in today's East Asian technology companies. The first case focuses on Kazuo Inamori (1932–2022), a Confucian entrepreneur who founded Kyocera in Japan, a world leader in advanced ceramic technology (Chou 2024b). Despite Japan's economic stagnation in the 1990s, Inamori's company continued to thrive. The second case highlights Morris Chang (1931–), who immigrated to the United States from China during his teenage years and later, in his fifties, founded TSMC in Taiwan (Miller 2022). Under his leadership, TSMC became one of the world's leading advanced semiconductor manufacturing companies. Together, these two cases offer valuable insights into how Confucian diligence shapes the success of contemporary East Asian high-tech companies.

4 Japan: Kazuo Inamori and the Kyocera Group

After World War II, Japan's economic and technological development was significantly influenced by both American policies and internal innovation (Miller 2022). Following the war, the United States, aiming to prevent Japan from rebuilding its military capabilities, imposed a pacifist constitution and facilitated the transfer of advanced technologies to Japanese scientists. This included crucial know-how such as the transistor technology, which played a pivotal role in Japan's subsequent technological advancements.

In the following decades, Japan experienced a remarkable rise in both its economy and technological capabilities (Miller 2022). Japanese automakers, such as Toyota, became renowned for producing high-quality vehicles at competitive prices, which led to their global success. The country's consumer electronics industry, exemplified by companies like Sony, achieved worldwide recognition for its creativity and superior product quality. Additionally, Japan's semiconductor industry also emerged as a global leader, with companies like NEC and Toshiba producing high-quality, cost-effective semiconductors that often surpassed American counterparts in performance and affordability.

However, Japan's rapid economic growth led to an overheated real estate market, driven by speculative investments (Miller 2022). The asset bubble burst in the early 1990s, exacerbated by the appreciation of the yen, which further strained the economy. This downturn resulted in a prolonged period of stagnation, known as the "Lost Decades." Not until the early 2020s did Japan begin to show signs of economic recovery and growth.

During Japan's economic bubble period, Inamori (2004), a distinguished entrepreneur who founded two Fortune 500 companies, Kyocera and KDDI, resisted the allure of the soaring real estate prices. Instead, he concentrated on enhancing his companies' competitive edges. As a result, Kyocera, a global leader in ceramic technology, continued to thrive despite Japan's subsequent economic stagnation. In 2000, KDDI became the second-largest player in Japan's telecom market, just behind NTT DOCOMO. In 2010, Inamori spearheaded the turnaround of the bankrupt Japan Airlines, which became the most profitable airline in the world in just three years (Chou 2024b).

Inamori deeply admired ancient Chinese sages and acknowledged the profound influence of Confucian thought on his work (Chou 2024a). He was widely regarded as a model of Confucian ethical leadership. This section aims to examine how he demonstrated Confucian diligence at work and how his work ethics influenced his companies' success.

4.1 Relentless Efforts

After graduating from college, Inamori (2017) secured his first job at Shofu Industries, a struggling company that frequently delayed employee paychecks. Nevertheless, Inamori chose to honor his duty as an engineer, believing that his destiny was determined more by his attitudes than by his surroundings. He understood that a complaining attitude in any environment would lead to misfortune, while a responsible approach could turn his fate around. From that point on, he committed tremendous effort to his work, often spending day and night in the laboratory. He even moved cooking tools like pots and pans into the lab to sustain himself and slept there while immersed in his projects. He also visited the library to consult academic journals, further enhancing his knowledge.

After leaving Shofu and founding Kyocera in 1959, Inamori (2017) maintained his relentless work ethic. He recognized that many Japanese companies had begun to make significant strides since the end of World War II in 1945, which meant that Kyocera was already 14 years behind its competitors. To close this gap, Inamori and his team dedicated themselves to working harder than anyone else in the industry. As a result, just twelve years after its founding, Kyocera went public. With this milestone, Inamori (2007) felt an even greater sense of responsibility, knowing that both employees and shareholders depended on his leadership. This awareness fueled his commitment to work even harder for the success of the company.

Inamori's (2009) philosophy of relentless effort was deeply influenced by his observations of nature. He noted that many plants in the tundra of the far north strive to flower and produce as many seeds as possible during the brief summer. Similarly, plants in the deserts of Africa quickly sprout buds and blooms and bear seeds after the rare rains that occur just once or twice a year. From these examples, Inamori realized that all living beings fully embrace each precious day without wasting time. He believed that we, too, should cherish every single moment and live wholeheartedly. Inamori (2017) posited that this earnest dedication to work not only defines our greatness but also fosters the development of a virtuous character. In contrast, those who shy away from hard work miss the opportunity to realize their true potential.

This relentless pursuit of excellence was key to Inamori's (2009) success in growing Kyocera from a modest startup into a Fortune 500 company. He acknowledged that at the outset, his ambition to transform the startup into the best in the world seemed bold. However, he came to realize that we shape our lives through the accumulation of our daily actions. As he stated, "Life is, after all, the sum of all our todays, the continuation of now. The accumulation of each second becomes a day; the accumulation of each day becomes a week, a month, a year" (p. 16). He concluded, "There is no shortcut to the destination we have painted in our minds. A great distance can be traveled only one step at a time, and a great dream is reached little by little, day by day" (p. 16).

Accordingly, Inamori emphasized that "repeated effort makes the ordinary extraordinary" (p. 15). He cited Kyocera's success as a prime example: "It was through steady, accumulated effort that Kyocera grew into what it is today and I went from an unknown engineer to a respected businessman. It is far better to devote your efforts to living fully every second of every day than to waste time worrying about tomorrow or focusing your thoughts on trying to read the future. In the end, this is the best way to realize your dreams" (p.17).

4.2 Ceaseless Improvement

Inamori emphasized that continuous effort does not necessarily equate to mere repetition of work. Instead, he advocated for a mindset of daily improvement, stating:

> Continuous effort is important, but this does not mean that people should merely do the same thing over and over. Continuity and repetition are different things. Rather than mindlessly repeating what we did the day before, we should seek ways to make tomorrow better than today, always adding some improvement tomorrow to our efforts from today. This creative attitude of innovation accelerates the rate at which we can achieve success. As an engineer, I am trained to ask myself if my results are good enough or whether there is a better way. If you look at what you are doing from that perspective, there is infinite potential for innovation in even the most mundane tasks (Inamori 2009, p. 17).
>
> Constantly striving to improve your work and make tomorrow even a little better than today and the day after that, better than tomorrow. If I was repeating the same research or the same task, I would try a different method. I kept my mind focused on ways to make my work even more efficient the next day. Using such creative ingenuity resulted in progress and developments that far exceeded what I had envisioned (Inamori 2017, Location 834).

Furthermore, Inamori (2014) noted that humans possess unlimited potential and that our abilities can continue to grow and evolve as we move toward the future. What seems unattainable today may well be achievable in a few years. Therefore, it is crucial to start learning now, embracing continuous improvement and refinement. By doing so, we can cultivate exceptional capabilities in the future. This principle of ceaseless improvement is essential for both personal and corporate growth. As stated by Inamori (2009, p.17), "Daily innovations lead to dramatic progress." Embracing this belief, he consistently achieved groundbreaking technological advancements.

For instance, during his first job, Inamori (2017, 2020) became the first person in Japan and the second one in the world to successfully mass-produce the ceramic U-shaped Kelcima for TVs and to synthesize highly pure forsterite in a laboratory setting. After founding Kyocera, Inamori maintained this spirit of ceaseless improvement. In Kyocera's early days, its potential clients were hesitant to work with startups unless they could meet demanding specifications that major suppliers refused to fulfill. Inamori decided to take on these daunting projects, rallying his team to develop solutions within tight deadlines and effectively achieving what seemed impossible (Chou 2024b).

Through this continuous process of self-challenge, progress, and innovation, Kyocera expanded its customer base from Panasonic to include Toshiba and Hitachi, steadily accumulating technological expertise and capabilities (Inamori 2014). Later, Inamori drew on Kyocera's past development projects to meet requests for components for Sony's tape recorders, benefiting from his company's growing reservoir of technology.

With this mindset, Kyocera made significant strides in technology development, ultimately establishing itself as a leading technology company in its field (Inamori 2017). Nowadays, Kyocera's fine ceramics have become indispensable industrial materials used across a wide range of fields, particularly those requiring advanced technology. Its ceramic technology, characterized by superior strength, corrosion resistance, heat tolerance, and insulation properties, has been employed in Japan's Hayabusa Asteroid Probe, the first mission to successfully retrieve material from an asteroid and bring it back to Earth. Additionally, its ceramic packages are utilized in the K computer, a renowned supercomputer that was once among the fastest in the world.

Inamori (2014) concluded that every creative idea may seem insignificant at first, but over time, it can lead to remarkable achievements. By consistently dedicating daily effort and making continuous improvements, companies can eventually foster significant innovations and technological advancements.

4.3 Boundless Passion

In addition to making relentless efforts at work for ceaseless improvement, Inamori underscored the importance of cultivating genuine passion for one's job. He stated, "To keep striving every day, you must learn to fall in love with it" (Inamori 2017, Location 417). He elaborated on this idea as follows:

> Love is the greatest motivator, the mother of enthusiasm, and effort is the road to success. This concept is expressed best by the Japanese proverbs 'A thousand miles seem like one when traveling for love' and 'Love leads to mastery.' When we love what we do, we naturally feel enthusiastic about it and want to give it our best effort, which consequently helps us improve very quickly. We even begin to look forward to doing things that others see as the most tedious chores (Inamori 2009, p. 41).

Inamori (2009) learned to love his work during his first job. As previously mentioned, he committed himself to his role as an engineer by putting all his efforts into developing new technologies for products. To his surprise, once he dedicated himself to his work, he achieved remarkable results, which made his job more interesting and sparked even greater passion for his research. He came to the realization that love and devotion are mutually reinforcing; when we love our work, we are more inclined to devote ourselves to it, and conversely, when we devote ourselves to our work, we grow to love it. Thus, he articulated the significance of passion for one's work:

> Working requires tremendous energy, an energy that arises from getting fired up. The best way to get fired up is to love your work. No matter what your job is, if you put your heart and soul into it, you will gain satisfaction and self-confidence and want to set out to achieve the next goal. As you repeat this process, achieving goal after goal, you will come to love your work even more, and eventually, no effort will seem painful and you will be able to achieve wonderful results (Inamori 2009, p. 41).

Inamori (2017, Location 417) concluded, "People who love their work and find joy in it can achieve success." Fueled by boundless passion and relentless effort toward ceaseless improvement, he grew his companies from obscure startups into Fortune 500 enterprises through continuous technological advancements. He also transformed from a boy from a rural area into a world-class entrepreneur, revered by the Japanese as "the sage of management in Japan" (Chou 2024a). Evidently, as a Confucian ethical leader in the modern technology business, Inamori exemplified the spirit of Confucian diligence.

In summary, while the goal of becoming a sage-king and that of advancing technology may seem distinct, they share a commonality—both are challenging and elusive, requiring unwavering dedication, constant refinement, and endless enthusiasm. Therefore, Inamori's diligent approach to building modern technology companies closely parallels the ancient Confucian scholars' pursuit of the sage-king ideal. Both paths embody three essential elements: relentless effort, continuous improvement, and boundless passion.

When these three elements are present, technology companies are more likely to thrive in competitive markets. Thus, incorporating Confucian diligence into modern technology management fosters rapid advancements in East Asia, especially when supported by factors such as reliable legal frameworks and adequate capital. By reinterpreting Confucian diligence in this context, we gain valuable insights into why many East Asian countries excel not only in labor-intensive manufacturing but also in technological innovation.

5 Taiwan: Morris Chang and TSMC

Following the exploration of Inamori's approach to diligence, the focus of this chapter shifts to TSMC. This section will examine whether TSMC and its founder exemplify Confucian diligence and how this ethical value has contributed to the company's achievements in the highly competitive semiconductor industry.

Contrary to the perception of Taiwan as a low-tech manufacturing hub in the 1980s, TSMC's current achievements in the semiconductor industry highlights a remarkable transformation that would have once seemed unimaginable. In fact, as early as the 1960s, East Asia began to forge connections with the Western semiconductor industry, with both sides actively engaging in collaboration and development (Miller 2022). In the 1960s, U.S. chip makers including Fairchild, Texas Instruments (TI), and Motorola began relocating their semiconductor assembly lines to East Asia, including Hong Kong, Taiwan, and Singapore, drawn by the region's efficient and low-cost labor. In parallel, the U.S. government expected this shift to provide stable jobs for millions of workers in East Asia and mitigate the spread of communism. Meanwhile, as China expanded its military capabilities, including nuclear weapons, the Taiwanese government sought to bolster national security by strengthening economic ties with the U.S. through the development of its semiconductor industry. Consequently, the Taiwanese government approved TI's plan to establish semiconductor assembly factories in 1968 and invited RCA to set up manufacturing facilities in 1976.

In the 1980s, the Taiwanese government invited Morris Chang, a former TI executive, to build semiconductor manufacturing factories in Taiwan, leading to the establishment of TSMC in 1987. In 1988, Intel's then CEO Andrew Grove visited Taiwan to assess the nascent chipmaker (Chen and Huang 2021). Intel engineers presented TSMC with over 200 complex problems to solve over the year. Chang emphasized to his team that producing chips for Intel would significantly enhance TSMC's engineering capabilities. TSMC engineers methodically tackled each challenge, ultimately securing their first order from Intel.

Three decades later, the roles shifted dramatically. In 2018, TSMC launched commercial production of the 7-nm process, surpassing Intel as the global leader in semiconductor manufacturing technology (Miller 2022). By 2024, TSMC is responsible for approximately 90% of the world's most advanced semiconductor chips (Kristof 2024). TSMC has grown from a fledgling startup in 1987 to become the world's largest dedicated independent semiconductor foundry.

TSMC's remarkable success is widely attributed to its workers' diligence. For instance, a U.S. engineer recruited by TSMC shared his observations during his 18-month training in Taiwan starting in 2021 (Liu 2024). His first impression was, "Oh, my gosh, people work hard." He concluded that this is what it takes to succeed in the semiconductor industry. Similarly, a BBC journalist noted that TSMC's achievements are difficult to replicate, as the company's secret to success has been honed through "decades of laborious work by its engineers" (Wingfield-Hayes 2023). In the context of the geopolitical tensions between the U.S. and China, Dr. Shih, a key figure in the establishment of TSMC, remarked, "I hope people appreciate the previous effort that we made and won't destroy it" (Wingfield-Hayes 2023). Phrases such as "work hard," "laborious work," and "the previous effort" emphasize the critical role that a strong work ethic, specifically diligence, plays in TSMC's accomplishments. Therefore, this section will analyze Morris Chang's life story and philosophy, along with TSMC's corporate culture, within the framework of this diligent work ethic.

5.1 Morris Chang and His Semiconductor Dream

Morris Chang, the founder of TSMC, is a pivotal figure in the global semiconductor industry. After immigrating from China to the U.S. in his youth, he earned a B.S. and M.S. in mechanical engineering from MIT and a Ph.D. in electrical engineering from Stanford. Early in his career at TI, he faced the challenge of improving a production line for unreliable transistors with nearly zero yield (the percentage of good output versus total output). Within months, he increased the yield on this line to 25%, showcasing his exceptional ability and commitment to *continuous improvement*. He soon assumed control of TI's entire integrated circuits (IC) business (Miller 2022).

Chang has always been deeply *passionate* about semiconductor manufacturing. In the early days, semiconductor companies like TI, Motorola, and Intel managed both their own design teams and manufacturing facilities. By the mid-1970s, Chang envisioned a groundbreaking business model: a company that would focus exclusively on manufacturing (i.e., a foundry), allowing semiconductor startups to specialize in IC design (fabless design houses). However, TI did not embrace his idea. Later, he was excluded from TI's succession plan and relegated to less significant roles. This ultimately prompted his departure from TI in the early 1980s (Miller 2022).

Despite leaving TI, Chang remained committed to his foundry vision. His business plan eventually gained the support of the Taiwanese government, which promised to provide 48% of the funding, stipulating that he find a foreign chip company to supply advanced production technology (Miller 2022). After contacting more than a dozen Western and Japanese semiconductor firms—including Intel, Mitsubishi, and Toshiba—Chang faced rejection from all (Business Weekly 2023). However, his *relentless efforts* bore fruit when Philips from the Netherlands agreed to transfer technology and invest 27.5% of the funding. The remaining shares were acquired by Taiwanese conglomerates (Miller 2022). This innovative foundry model enabled the rise of the fabless IC industry; for instance, the success of Nvidia, a leading U.S. AI chip company, would not have been possible without TSMC's advanced technology and business model.

Chang's diligence, evident in his relentless effort, continuous improvement, and boundless passion, not only led to the creation of TSMC but also established it as the world's leading foundry. Chang is deeply dedicated to the relentless pursuit of improvements in semiconductor manufacturing technologies, which require significant investment even during downturns. For example, in response to the 2008 financial crisis, Rick Tsai, Chang's handpicked successor, laid off employees and cut costs, as most CEOs would. However, Chang recognized that it was crucial to maintain research investments to enhance TSMC's technological competitiveness (Miller 2022). As a result, he stepped back into the role of CEO in 2009, rehiring the engineers laid off by his successor and doubling down on research and development investments to uphold his commitment to technological advancement during challenging economic times.

5.2 Morris Chang's Life Philosophy

Chang's work ethic of diligence is deeply rooted in his life philosophy, which revolves around lifelong learning for continuous improvement. He views learning as a precious gift that many people neglect despite their innate curiosity (Business Weekly 2023). Chang dedicates himself to daily learning, making it a central part of his life. He commits to reading at least six hours a day: three hours at work, focusing on semiconductor-related materials and management knowledge, and another three hours at home, exploring unrelated topics such as economics, politics, and history (Chen 2015).

Chang also stresses the importance of integrating learning with critical thinking (Business Weekly 2023). He asserts that without independent thought, lifelong learning lacks effectiveness; without continuous learning, there is little substance to reflect on. His perspective mirrors Confucius's wisdom: "To study and not think is a waste. To think and not study is dangerous" (*Analects* 2:15; Muller 2021). This combination of relentless learning and critical thinking requires a commitment to continuous effort, a desire for improvement, and a passion for both knowledge and reflection—qualities that perfectly embody the spirit of Confucian diligence.

In addition to lifelong learning, Chang highlights the significance of passion in one's work (Business Weekly 2023). He believes that rigid career planning can hinder commitment and engagement in current roles. Instead, Chang advocates focusing on personal interests and making meaningful contributions in every position. He emphasizes that the best career strategy is to pursue what truly interests us at each stage of our careers while consistently striving to contribute to the company's success. This philosophy, blending dedication with passion, aligns seamlessly with the values of Confucian diligence.

In essence, Morris Chang's life philosophy—anchored in lifelong learning, critical thinking, and a passion for one's work—demonstrates how Confucian diligence can lead to both personal growth and professional success. Similarly, this spirit of Confucian diligence also plays a key role in driving TSMC's corporate success.

5.3 TSMC's Corporate Culture

Business studies often assert that the personalities of founders shape the corporate culture of their firms. In the case of TSMC, Morris Chang's work ethic of diligence not only influences the company's corporate culture but also resonates with the values of the Taiwanese people, both of which contribute to TSMC's world-class success.

5.3.1 Relentless Effort

In a BBC report (Wingfield-Hayes 2023), a Taiwanese engineer, speaking on condition of anonymity, emphasized the labor advantage of Taiwanese companies. Using himself as an example, the engineer shared that he works six days a week, starting at 7:30 a.m. and typically finishing at 7:00 p.m. If issues arise at the plant outside regular work hours, he is often called in to resolve them, even on Sundays or holidays.

TSMC employees demonstrate even greater dedication, which has astonished many observers. Shang-Yi Chiang, a former TSMC executive, compared his experiences in the U.S. with his observations at TSMC, stating, "People work so much harder in Taiwan" (Miller 2022, p. 232). He noted that if equipment breaks down at 1 a.m. in the U.S., engineers would typically address it in the morning. In contrast, at TSMC, the issue is often resolved by 2 a.m. TSMC engineers do not complain about these demands, nor do their spouses; it is not uncommon to find engineers willing to work all night in Taiwan.

This phenomenon, though unusual, reflects a profound commitment to hard work, deeply rooted in the Confucian value of diligence. However, this relentless attitude has also caused cultural clashes during the construction of TSMC's plants in the U.S. (Liu 2024). While it underscores the extreme application of Confucian diligence, it also highlights the challenges of integrating such a work ethic into different cultural contexts.

5.3.2 Ceaseless Improvement

Success in the foundry business requires not only technological breakthroughs but also high yields, which directly affect profit margins. The fabrication of a semiconductor chip involves thousands of engineering steps, and the quality of each step significantly impacts the company's financial results. For example, if it takes a thousand steps to build a chip and each step achieves a yield of 99.99%, the final yield will be 90.48% (calculated as $99.99\%^{1000} = 90.48\%$). If each step's yield is 99.90%, which appears to be only a slight decrease, the final yield drops dramatically to 36.77% ($99.90\%^{1000} = 36.77\%$). In this example, a slight 0.01% difference in yield at each step can lead to a substantial 53.71% variation in the final yield after a thousand manufacturing steps. While a 90.48% yield can significantly boost corporate profits, 36.77% is likely to result in substantial financial losses.

Given this context, it is crucial for foundry businesses to patiently improve every minute detail that contributes to the final yield. Engineers at TSMC continuously calibrate manufacturing equipment and make small adjustments through trial and error to enhance yield. The BBC has drawn a parallel between cooking and semiconductor manufacturing: with the same ingredients and recipes, a more skilled cook can create a superior dish (Wingfield-Hayes 2023). For TSMC, the "secret sauce" lies in its unwavering commitment to ceaseless improvement.

5.3.3 Boundless Passion

Based on Shang-Yi Chiang's observation, Morris Chang's ambition to develop world-class semiconductor technology permeates the entire culture of TSMC. This passion plays a crucial role in the relentless efforts of TSMC engineers to pursue continuous improvement (Miller 2022). For instance, although some TSMC's Taiwanese engineers at its Arizona site share their American colleagues' frustrations with strict discipline and long hours, they continue to work diligently and passionately. As one TSMC engineer stated anonymously, "Everything comes from working hard. Without this culture, TSMC cannot be number one in the world... I want to support TSMC to be great. It's my religion" (Zhou 2024).

To inspire passion for work, a TSMC manager emphasizes the importance of viewing one's work as a "work of art" rather than just a job (Yu 2019). He states that individuals are more willing to invest time and effort into perfecting their art, an attitude that contrasts sharply with simply fulfilling job duties. For instance, an equipment engineer who approaches his work as a work of art will go beyond basic requirements, such as ensuring that the screws on a robotic arm are tightened and the machine operates smoothly. Instead, he will seek innovative solutions—considering precisely how many turns are optimal for tightening the screws and how to achieve more accurate positioning of the robotic arm. In other words, treating work as a work of art can ignite enthusiasm, driving workers to pursue relentless effort and ceaseless improvement.

In summary, Chang and TSMC employees exemplify a relentless pursuit of excellence, ceaseless improvement in technology, and boundless passion for advancing semiconductor manufacturing technology, embodying the spirit of Confucian diligence. Their strong work ethic has yielded significant results: Under Chang's leadership, TSMC has not only revolutionized the manufacturing of integrated circuits but has also established a benchmark for innovation and quality within the industry.

6 Discussion

In light of the inadequacy of past interpretations of Confucian work ethics in explaining the current landscape of technological development in East Asia, this paper conducts an in-depth analysis of Confucian diligence. This section explores the theoretical contributions of this article, highlight the implications of the findings for various stakeholders, address the research limitations, and outline potential future directions that can further enrich our understanding of the intricate relationship between culture and technological progress in East Asia.

6.1 Theoretical Contributions

This article makes three contributions to the understanding of Confucian diligence. First, it detects the gap in Redding's interpretation of Confucian diligence. Second, it elucidates the concept of Confucian diligence. Lastly, it clarifies the need for cultural-system analysis to enhance our understanding of this cultural construct.

6.1.1 Identifying Gaps in Redding's Interpretation of Confucian Diligence

The East Asian region has been profoundly influenced by Confucian values, leading scholars to interpret its economic development through the lens of Confucianism. For instance, in the early twentieth century, Weber (1951) suggested that the lack of inner-worldly asceticism in Confucian thought contributed to China's inability to develop modern capitalism independently. Conversely, towards the end of the twentieth century, Redding (1993) utilized the Confucian work ethic to explain the region's thriving labor-intensive industries, emphasizing that the diligence instilled by Confucianism enables individuals in East Asia to endure monotonous, repetitive tasks. While Confucianism has undeniably been a crucial cultural force in East Asia's formerly low-tech economy, the rapid technological advancements in Taiwan, South Korea, and China over the past three decades—mirroring Japan's achievements in the 1980s—suggest that Redding's definition of Confucian diligence is insufficient to fully explain the region's current technological progress and economic development.

6.1.2 Elucidating the Construct of Confucian Diligence

Accordingly, this paper examines how the concept of Confucian diligence transcends Redding's interpretation by integrating relentless effort, ceaseless improvement, and boundless passion. Rooted in the ancient sage-king model and Confucian scholars' ideals, these elements are pivotal not only in cultivating a work ethic that spans generations but also in shaping contemporary technological development in East Asia. The success of Japan's technological achievements in the 1980s, as well as the rapid advancements in Taiwan, South Korea, and China over the past three decades, can all be attributed to Confucian diligence.

This work ethic is not limited to high-tech sectors but permeates various industries across East Asia. For instance, in a 2024 TV interview, Jensen Huang, co-founder and CEO of Nvidia, expressed his admiration for the hardworking and passionate attitudes of workers at Taiwanese night markets—vibrant street venues known for their diversity in culinary offerings (TVBS 2024). Huang's observation underscores how Confucian diligence shapes a strong professional ethic that fuels economic development across a wide array of fields in East Asia.

6.1.3 Highlight the Importance of Cultural-System Analysis

Finally, the manifestation of culture is a complex phenomenon that evolves across varying temporal and spatial contexts. The enduring values within a cultural system may not always be fully expressed at certain times or in particular environments. For example, while Japan fully exemplified the Confucian work ethic of diligence—encompassing effort, improvement, and passion—during the 1980s, the Four Asian Tigers were still perceived in the 1990s primarily through the lens of perseverance in repetitive and monotonous work (e.g., Redding 1993). Over the next three decades, technology companies in Taiwan, South Korea, and China further embodied ceaseless improvement and boundless passion, driving significant advancements in technology and reshaping the global innovation landscape. This suggests that cultural analysis should not be limited to observable phenomena but should also delve deeper into the underlying cultural systems that shape socio-cultural interactions.

By incorporating cultural-system analysis alongside empirical studies, scholars can avoid the pitfall of research being overly driven by observable trends. This dual approach allows for a deeper exploration of the underlying cultural values that shape visible behaviors and practices. While empirical studies provide valuable data on present phenomena, cultural-system analysis uncovers the historical and contextual factors that give rise to these trends. This comprehensive approach leads to more nuanced interpretations of cultural practices, such as the Confucian work ethic, and their broader implications for contemporary issues like technological advancement and economic development. Ultimately, integrating these methodologies enriches the discussion of cultural influences, enabling researchers to draw more meaningful conclusions that go beyond surface-level observations.

6.2 Practical Implications

The findings of this paper emphasize the significant role of cultural values, particularly Confucian diligence, in shaping the economic and technological development of East Asia. This underscores the necessity for both corporate leaders and policymakers to understand the cultural contexts that influence technology innovation and business success in these regions. For corporate leaders, embodying the qualities of boundless passion, relentless effort, and ceaseless improvement is essential for fostering a productive work environment and driving organizational performance. By promoting a culture that values these attributes, companies can enhance their competitive edge in the market. Similarly, policymakers should integrate cultural considerations into their frameworks for promoting technological innovation, recognizing how these values impact educational and business practices. This comprehensive understanding can help create an environment conducive to sustained growth and advancement.

Nevertheless, it is noteworthy that while this article aims to elucidate the cultural factors underlying technological development in East Asia, it does not make a definitive claim that East Asian technological capabilities surpass those of the West. As early as the 1980s, Japan highlighted that while the United States excelled at *disruptive* innovation, Japan was better at *continuous* innovation (Miller 2022). Therefore, it may be more accurate to say that this paper seeks to explain the cultural factors behind East Asia's achievements in continuous innovation, particularly focusing on the work ethic derived from Confucian diligence.

However, it may be overly simplistic to attribute technological success in East Asia solely to continuous innovation. For instance, since the 1970s, Sony has launched revolutionary consumer electronics products, such as the Walkman. Similarly, to enhance its competitive advantages, TSMC consistently pushes the envelope of semiconductor process technologies through disruptive innovations, including advancements in copper process technology, immersion lithography, 3D packaging, and FinFET transistors (Business Weekly 2023; Hsiao 2024; Hwang 2024; Patterson 2023). In fairness, different technology sectors require different strategies to stay competitive. While many other technology products may rely primarily on disruptive innovation for success, excellence in the semiconductor manufacturing technology requires a combination of both disruptive and continuous innovations.

6.3 Research Limitations and Future Directions

In exploring the cultural factors that shape technological development in East Asia, this paper acknowledges certain limitations that warrant further discussion. Addressing these limitations can pave the way for future research that enriches our understanding of the complex dynamics between culture and technological advancement.

6.3.1 Need for Empirical Research and Cultural Comparison Analyses

This paper examines the concept of Confucian diligence and its impact on technological development from a theoretical perspective. Future research could benefit from qualitative and quantitative research to validate this construct and examine its antecedents and consequences. Such efforts would deepen our understanding of this cultural concept.

Future studies could also compare the notion of diligence in other cultures with Confucian diligence. For example, the work ethic of diligence associated with Calvinism serves to honor God and assess one's status as the elect, which likely differs from the Confucian conception of diligence that evolves from the sage-king tradition. This presents an area worthy of further exploration.

6.3.2 Need for Discussion on the Normative Issue of Work-Life Balance

From a Western perspective, Confucian diligence may appear to conflict with the value of work-life balance, presenting a normative issue that warrants further scholarly investigation. While this paper primarily focuses on clarifying the concept of Confucian diligence and its influence on technological development in East Asia, it does not directly address the work-life balance dilemma. Future normative studies could explore the tension between maintaining a balanced life and the demands of diligent work.

The concept of free will could play a key role in this discussion. If individuals freely choose to work hard, even at the expense of work-life balance, such decisions may deserve respect. However, when relentless work is imposed rather than chosen, it should not be encouraged. It is worth noting that the passion embedded in Confucian diligence reflects a work ethic rooted in self-determination. Yet in practice, it remains important to examine whether East Asian engineers share the same level of passion as their leaders.

Looking back in time, over three thousand years ago, long before East Asia's technological prosperity, King Wen of Zhou, a benevolent sage-king, wrote these words: "Just as wood grows from beneath the earth and eventually rises, a gentleman follows moral principles, accumulating small virtues and thus achieving great things" (*I-Ching* 46). Even while imprisoned by the Shang tyrant, King Wen emphasized that the cultivation of virtue, much like the growth of a tree from a tiny seed, requires passionate effort and steady, uninterrupted progress toward greatness. This enduring spirit of diligence not only inspired Confucius and Mencius in their pursuit of the sage-king ideal but also continues to drive modern East Asia's study and work ethics.

For example, research in cultural psychology shows that Taiwanese education consistently instills the value of diligence. Parents and teachers prioritize effort and perseverance, offering support to children who struggle academically but continue to work hard (Chen 2023; Chen et al. 2016). This emphasis on persistence extends into professional life, where individuals apply the same dedication to their careers, reinforcing the Confucian work ethic that fuels the region's rapid economic and technological advancements. Ultimately, the timeless values embodied by King Wen of Zhou remain deeply ingrained in East Asian culture, shaping not only personal virtue but also collective economic progress of the region.

References

Archer, M. S. (1995). *Realist social theory: The morphogenetic approach.* Cambridge: Cambridge University Press.

Archer, M. S. (2005). Structure, culture and agency. In M. D. Jacobs & N. W. Hanrahan (Eds.), *The Blackwell companion to the sociology of culture* (pp. 17–34). Oxford: Blackwell Publishing.

Baillie, M. G. L., & McAneney, J. (2014). Why we shouldn't ignore the mid-24th century BC when discussing the 2200–2000 BC climate anomaly. 7th Archaeological Conference of Central Germany, Halle, Germany.

Barro, R. J. (1998). *The East Asian Tigers have plenty to roar about.* Bloomberg. https://www.bloomberg.com/news/articles/1998-04-26/the-east-asian-tigers-have-plenty-to-roar-about?embedded-checkout=true. Accessed 10 August 2024.

Business Weekly (2023). *The way of TSMC: The philosophy and strategy of Morris C. M. Chang to make TSMC ahead of the market* (in Chinese). Taipei: Business Weekly.

Chan, W.-T. (1963). *A source book in Chinese philosophy.* Princeton University Press.

Chen, F. (2015). *You don't need to read the entire book* (in Chinese). Manager Today. https://www.managertoday.com.tw/articles/view/51510. Accessed 10 August 2024.

Chen, S.-W. (2023). Learning motivations and effort beliefs in Confucian cultural context: A dual-mode theoretical framework of achievement goal. *Frontiers in Psychology, 14,* 1058456.

Chen, L.-R., & Huang, E. (2021). *The coming TSMC boom: Can it double sales by 2023?* CommonWealth Magazine. https://english.cw.com.tw/article/article.action?id=2928. Accessed 10 August 2024.

Chen, S.-W., Fwu, B.-J., Wei, C.-F., & Wang, H.-H. (2016). High-school teachers' beliefs about effort and their attitudes toward struggling and smart students in a Confucian society. *Frontiers in Psychology*, *7*, 1366.

Chou, S. C.-M. (2024a). Beyond reach but within sight: Ethical leaders' pursuit of seemingly unattainable role models in East Asia. *Journal of Business Ethics, 195*, 631–652. https://doi.org/10.1007/s10551-024-05659-2.

Chou, S. C.-M. (2024b). The journey from unrealistic to realistic optimism: Ancient Confucian wisdom in modern business practice. *Culture & Psychology.* https://doi.org/10.1177/1354067x241232109.

Hsiao, M. (2024). *Interview with the father of immersion lithography who put TSMC on the map.* CommonWealth Magazine. https://english.cw.com.tw/article/article.action?id=3720. Accessed 10 August 2024.

Hwang, K.-K. (2015a). Cultural system vs. pan-cultural dimensions: Philosophical reflection on approaches for indigenous psychology. *Journal for the Theory of Social Behaviour, 45,* 1–24.

Hwang, K.-K. (2015b). Culture-Inclusive theories of self and social interaction: The approach of multiple philosophical paradigms. *Journal for the Theory of Social Behaviour*, *45*(1), 40–63.

Hwang, J.-S. (2024). *Samsung to unveil 3D AI chip packaging tech SAINT to rival TSMC.* The Korea Economic Daily. https://www.kedglobal.com/korean-chipmakers/newsView/ked202311120002. Accessed 10 August 2024.

Inamori, K. (2004). *Kazuo Inamori's autobiography* (in Japanese). Tokyo: Nikkei Publishing.

Inamori, K. (2007). *The kingly way of life* (in Japanese). Tokyo: Nikkei Business Publications.

Inamori, K. (2009). *A compass to fulfillment: Passion and spirituality in life and business.* New York: McGraw Hill.

Inamori, K. (2014). *Kyocera philosophy* (in Japanese). Tokyo: Sunmark Publishing.

Inamori, K. (2017). *A compass to wonderful life*. Tokyo: PHP Institute.
Inamori, K. (2020). *From zero to Kyocera*. Tokyo: PHP Institute.
Kristof, N. (2024). *Visiting the most important company in the world*. The New York Times. https://www.nytimes.com/2024/01/24/opinion/tsmc-taiwan-china.html.
Legge, J. (1861a). *Da Xue*. Chinese Text Project. https://ctext.org/liji/da-xue. Accessed 10 August 2024.
Legge, J. (1861b). *The Analects*. Chinese Text Project. https://ctext.org/analects. Accessed 10 August 2024.
Legge, J. (1895). *Mengzi*. Chinese Text Project. https://ctext.org/mengzi. Accessed 10 August 2024.
Liu, J. (2024). *Why TSMC has struggled to manufacture chips in Arizona*. The New York Times. https://www.nytimes.com/2024/08/08/business/tsmc-phoenix-arizona-semiconductor.html. Accessed 10 August 2024.
Mathai, K., Gottlieb, G., Hong, G. H., Jung, S. E., Schmittmann, J., & Yu, J. (2016). *China's changing trade and the implications for the CLMV economies*. International Monetary Fund. https://www.imf.org/external/pubs/ft/dp/2016/apd1601.pdf. Accessed 10 August 2024.
Miller, C. (2022). *Chip war: The fight for the world's most critical technology*. New York: Simon and Schuster.
Muller, A. C. (2021). *The Analects of Confucius*. Resources for East Asian Language and Thought. http://www.acmuller.net/con-dao/analects.html. Accessed 10 August 2024.
Patterson, A. (2023). *Samsung seen stumbling at silicon's leading edge*. EE Times. https://www.eetimes.com/samsung-seen-stumbling-at-silicons-leading-edge/. Accessed 10 August 2024.
Pines, Y. (2023). The Confucian legendary past. In J. Oldstone-Moore (Ed.), *The Oxford handbook of Confucianism* (pp. 57–69). New York: Oxford University Press.
Prescott, A. (2015). Introducing East Asia. In A. Prescott (Ed.), *East Asia in the world: An introduction* (pp. 3–24). New York: Routledge.
Redding, G. (1993). *The spirit of Chinese capitalism*. New York: Walter de Gruyter.
Richey, J. L. (2015). Society and culture: Confucianism in East Asia today. In A. Prescott (Ed.), *East Asia in the world: An introduction* (pp. 174–184). New York: Routledge.
TVBS (2024, June 16). *Exclusive interview with Nvidia CEO Jensen Huang*. Youtube. https://www.youtube.com/watch?v=IkFgiDy_578. Accessed 10 August 2024.
Weber, M. (1951). *The religion of China: Confucianism and Taoism*. Glencoe, IL: The Free Press.
Weber, M. (1981). *General economic history*. New Brunswick, NJ: Transaction Books.
Weber, M. (2001). *The Protestant ethic and the spirit of capitalism*. New York: Routledge.
Wingfield-Hayes, R. (2023). *The secret sauce for Taiwan's chip superstardom*. BBC. https://www.bbc.com/news/world-asia-67213293. Accessed 10 August 2024.
Yu, Y.-F. (2019). *The DNA of TSMC* (in Chinese). Taipei: Commonwealth Publishing.
Yü, Y.-S. (2021). *The religious ethic and mercantile spirit in early modern China*. New York: Columbia University Press.
Zhou, V. (2024). *TSMC's debacle in the American desert*. Rest of World. https://restofworld.org/2024/tsmc-arizona-expansion/. Accessed 10 August 2024.

The Ethics of Digital Nudging

Cristiana Cerqueira Leal and Benilde Oliveira

Abstract The ethical debate on nudging often goes beyond the strict definition of nudging to encompass broader forms of influence and manipulation that are emerging with the rise of digital environments. The growing role of artificial intelligence, with its ability to personalize experiences and enable immediate interaction, raises new concerns about the power and reach of behavioral interventions, blurring the lines between ethical guidance and manipulation. This chapter presents nudging, its application in digital environments, the rise of hypernudging through artificial intelligence, and the ethical debates accompanying these practices. It examines the ethical foundations of freedom of choice, tracing their implications in the context of digital nudging and hypernudging environments. Moreover, it analyzes the legal framework for digital nudging and discusses the processes and outcomes of ethical nudging. Finally, it outlines the main principles guiding ethical nudges and proposes a new agenda for responsible nudging, in which organizations disclose responsible nudging practices and through which complying organizations would obtain responsible nudging certification.

Keywords Ethical nudging · Digital nudging · Hypernudging · Responsible nudging · Artificial intelligence (AI)

C. C. Leal (✉)
Centre for Research in Economics and Management (NIPE) & School of Economics, Management and Political Science, University of Minho, Braga, Portugal
e-mail: ccerqueira@eeg.uminho.pt

B. Oliveira
School of Economics, Management and Political Science, University of Minho, Braga, Portugal

C. Machado (ed.), *Ethics in Management and Business*, Management and Industrial Engineering, https://doi.org/10.1007/978-3-031-82054-0_3

1 Introduction: Nudges, Digital Nudges and Hypernudges

1.1 Nudges

In the nudging ethical discussion, clearly defining what nudging is and is not becomes fundamental. Since Thaler and Sunstein (2008) introduced the term, it has attracted significant positive attention, along with some ethical concerns about its effects. In our view, many of the ethical issues that have been raised may have been inflated, deriving from some misinterpretations of its core concept and scope.

A nudge is "*any aspect of the choice architecture that alters people's behavior in a predictable way without forbidding any options or significantly changing their economic incentives*" (Thaler and Sunstein 2008, p. 6). Moreover, to be considered a nudge, it must be easy and cheap to avoid, implying that all options remain available and can be obtained without significant cost or effort.

Inherent to the concept of nudge is the idea that the intervention is intended to improve decisions, though this is not explicitly stated in its strict definition. The subtitle of Thaler and Sunstein's 2008 book, *"Improving Decisions About Health, Wealth, and Happiness,"* implies it, and their entire reasoning centers around enhancing individual and societal well-being. However, this knowledge can be used in both directions: to improve or deteriorate the decision processes and outcomes.

As we live in the age of influence—given the constant influence exerted by various means in our daily lives—it is natural that the ethical debate surrounding nudging has attracted considerable interest, debating both the nudging process and the nudging outcomes. Frequently, nudges are also referred to as **soft libertarian paternalism** (Sunstein 2014), raising ethical concerns about autonomy. Autonomy discussion reflects the tension between guiding individuals toward beneficial choices and preserving their freedom to make decisions independently. Just because they are subtle or soft forms of influence does not mean these concerns should be minimized. Their subtlety or softness can eventually lead to more harmful effects, particularly if this knowledge is used to undermine decision-making processes rather than improve them.

To contribute to this distinction in the intent of intervention, Thaler (2018) proposes the alternative term when the purpose of intervention is to cause harm: **sludge**. Sludges can take two forms: they can prevent people from doing things in their best interest or push them toward detrimental actions. This highlights the idea that conscientious choice architecture aims to help people make better choices "***as judged by themselves***" (Thaler 2018; Sunstein 2018). Even in this domain, ethical discussion is still needed, particularly about forming preferences under the guise of being judged as by themselves. Hence, since the emergence of the concept of nudging, critics quickly arose to focus on the ethical implications of its application for changing preferences and the potential for manipulation (see, for instance, Bovens 2009; Hansen and Jespersen 2013).

1.2 Digital Nudges and Hypernudges

Nudging interventions that target a specific population are often implemented through a homogeneous method, assuming "***as judged by themselves***" as homogenous preferences. As a result, both the intervention process and the intended outcome tend to be standardized. These nudging interventions can be effective in the aggregate, especially when targeting universal goals such as health, longevity, or better quality of life. While the individual nudging impact may vary from person to person, the intervention is successful if the overall impact is positive. More personalized approaches, tailored to individual characteristics, are likely more effective than the one-size-fits-all method typically used in traditional nudging interventions.

Digital nudging enables personalized interventions based on individuals' characteristics, offering a more tailored approach than traditional nudging methods. **Digital nudging** refers to nudging interventions delivered through digital platforms, such as websites, mobile apps, and other online tools that track user behavior or require registration, thereby collecting individual information. Digital environments enable the collection of various personal data, such as demographic information, geographic location, and individual preferences. This data enables segmentation based on relevant characteristics, making nudges highly customizable, adaptable, and personalized. This level of personalization tends to enhance their effectiveness, as the interventions are tailored to the individual, leading to improved individual outcomes.

Digital nudging can be significantly advanced using big data and sophisticated computational processing systems. **Hypernudges** are big data analytical nudges (Yeung 2019). They enable a highly personalized choice architecture that is extremely powerful and effective. Dynamic nudging algorithms continuously adapt to influence more efficiently, acting as recommender systems. With the rise of artificial intelligence (AI) integrated into digital environments, the influence on decision-making processes is becoming increasingly profound. It facilitates the creation of highly personalized choice architectures that are powerful and effective due to their interconnected, constantly updated, and adapting characteristics. This results in an interactive process where the behavior of initial nudges is monitored, enabling real-time adjustments that improve their overall efficiency. The pervasive nature of these recommender systems ensures that nudges are consistently integrated into users' experiences. This hypernudging iterative process may evolve and reach the point where even initially well-intentioned nudging becomes invasive through the dynamic process of refinement and insistence on influence. The integration of personal data and the personalization of digital media improve diverse aspects of everyday life, operating as a significant aid for productivity and efficiency. However, this same technology can also become intrusive, repetitive, and exhausting, leading to reactance against what initially appeared to be a gift.

If nudging in its simplest forms raises ethical questions, hypernudging amplifies these concerns due to the inherent issues of privacy and the potential for more intrusive and repetitive forms of influence. Dynamic choice architecture in digital environments emerges through adaptive interfaces driven by big data analysis, which continuously adjust and personalize the user experience in real time (see, for instance, Leal and Oliveira 2024). These tools utilize vast amounts of data to refine decision pathways, increasing user engagement and tailoring choices to individual preferences based on past behaviors. Nudging personalization can reach levels of individualization for a vast set of specific characteristics and adapt to new information in real time.

Hypernudging through AI offers a remarkable opportunity to place the individual in the spotlight, with an immense capacity to accommodate their preferences, desires, and even emotions. However, it also raises severe ethical concerns about freedom, autonomy, and privacy, especially in an era where people are increasingly under constant observation and surveillance. By placing the individual at the center of a hyper-personalized choice environment, hypernudging can serve both as a tool for empowerment and as a mechanism for diminishing personal agency. The concern becomes even more pressing when decision-making power is subtly stripped away, all while creating a false sense of empowerment and control.

1.3 Can Non-nudging Be Less Ethical than Nudging?

A provocative question that challenges conventional thinking is whether inaction—i.e., choosing not to influence—can be less ethical than actively nudging. If evident knowledge exists that some option is a better option, could fail to intervene to point out that option be considered a form of **manipulation by omission**? Is non-nudging ethically preferable, or can it be a passive failure to decide not to act in the individual's best interest? In other words, suppressing influence that could guide individuals toward better outcomes can eventually be considered less ethical than the absence of that influence. Although the debate generally focuses on whether it is ethical to actively nudge, "*in many cases, nudges are ethically required, not forbidden*" (Sunstein 2016, p. 11).

In situations where knowledge and expertise are unevenly distributed, nudges may become required to rebalance powers in favor of the weaker side of this relationship. It cannot be considered reasonable to suppress beneficial nudging strategies, leaving decision-makers to decide under their recognized limited understanding, which would result in less informed and efficient decision-making processes and, ultimately, poorer outcomes.

When the choice architect holds a dominant position in terms of expertise in the decision-making process, aiming to design a neutral choice architecture, where all options are presented on an equal level, may lead to avoidable inefficiencies. The decision process would require more time, attention, and cognitive resources, potentially still resulting in poorer choices. For example, in financial decision-making, presenting all investment options equally—without highlighting some options or suggesting a recommended financial product based on risk tolerance—could overwhelm individuals, especially those with weaker financial literacy, leading to suboptimal investment decisions or even no decision at all due to inaction or procrastination.

While this may be labeled as soft paternalism, given the overwhelming number of decisions individuals face daily, simplifying some of those choices, especially in areas where their expertise is limited, can be not only useful but also very welcomed and ethically recommended.

1.4 The Inevitably of Nudging

In a theoretical scenario where nudging is seen as inherently bad, could it be possible to eliminate every choice architecture that possibly influences the decision-making process? In practice, the mere existence of choice implies a choice environment, even if that environment is entirely accidental and random.

The architecture of choice is inevitable (Sunstein 2015). Whether carefully thought out or thoughtless, a setting of choice must always be presented and inevitably framed. Even looking for transparent frames, framing will somehow exist. So, recognizing that the design of the choice environment can significantly shape decision outcomes is fundamental to approaching its construction in a reflective and deliberate manner, cautiously examining how it shapes choices. Consequently, the intention could be to minimize any conditioning, looking for transparent frames, or it could be to optimize conditioning strategies to promote better choices and improve individual well-being.

In digital environments, the architecture of choice becomes more pondered. Unlike in physical settings, where individuals might face varying choice architectures—whether intentional or accidental—the digital environment is deliberately designed from conception. Every scenario in the digital environment is by design and, consequently, premeditated, with specific objectives in mind, making the influence on decision-making more structured and purposeful.

In digital environments, the interaction starts with the platform presenting a specific setting. Its definition departs from carefully selecting the most effective environment for a specific condition. The presumed most straightforward, most engaging, or most dynamic formats are going to be the chosen ones. In this sense, framing is always present, as every aspect of the digital experience shapes how a situation is understood. From the wording used in descriptions to the visual elements, every detail is critical in framing a decisional context and influences the user's perception and decision making process. In digital environments, the design of choice is intricately embedded within the digital setting of nudging and hypernudging to produce choices by design. For better or worse, often claiming the best intentions in most cases, digital nudging and hypernudging may significantly impact decision-making and overall user satisfaction. In digital environments, it is possible to create profiles for different categories of users. In hypernudging the level of personalization of the choice environment can be taken to the extreme for every relevant characteristic of the individual in that decisional setting, being constantly updated.

The argument that choice architecture is inevitable and that people are constantly being influenced does not exempt intentional nudging interventions from ethical evaluation. Premeditated and informed influence produces significantly different outcomes than random influences, whether regarding the means employed or the intended outcomes. Therefore, nudges that use premeditated designs require careful evaluation.

2 Fundamental Ethical Concerns in Nudging

2.1 Freedom, Autonomy, and Non-interference

The concept of **freedom of choice** is central to discussing the ethics of nudging. While attempting to potentially improve outcomes, nudging might put individuals' freedom of choice in check. Even in the presence of well-intentioned purposes, the use of nudging faces the challenging balance between what is supposed to be the best for the individual, determined by an external entity considered to have the best knowledge of their interests, and the individual right to freely choose accordingly with her or his preferences, even if they are less rational or efficient.

Berlin's (1969) approach to the **liberty concept** offers a deeper understanding of the ethical implications of nudging by considering two types of liberty: positive liberty and negative liberty. **Negative liberty** refers to the absence of external interference, emphasizing the non-existence of any kind of coercion or external constraints. In contrast, **positive liberty** refers to autonomy and control over one's choices in the sense that the individual is self-governing in order to achieve self-realization.

From the perspective of **negative liberty (non-interference)**, nudging subtly influences the individual's autonomy by guiding decision-making without coercion by any means, preserving freedom of choice (especially when compared with other policy tools and regulations), and maintaining all options available. Defendants of nudging argue that negative liberty is respected as individuals are gently guided toward better decisions with their own best interests in mind while keeping all other options easily accessible and ensuring no coercion at any stage. However, a potential ethical conflict arises with regard to the level of influence on the decisions, even if no coercive mechanisms are involved. Thus, from the standpoint of negative liberty, the ethical assessment focuses on nudges as interference. Even if nudges are, by definition, non-coercive, they may subtly manipulate choices, restricting the total manifestation of a free choice. The ethical dilemma leans over how to ensure that nudges do not escalate into manipulative interference that harms genuine freedom of choice.

From the perspective of **positive liberty (autonomy)**, nudges can be understood as a mechanism for enhancing self-determination and facilitating the individual's self-realization, making, from the perspective of nudging defendants, nudging objectives more aligned with the positive liberty concept. For example, healthier financial habits such as retirement planning and saving help individuals achieve their goals and align with their long-term interests. Nudges help people make choices in their best interest and potentially enhance their ability to achieve their goals and live more fulfilling lives. Accordingly, they can be seen as a reinforcement of positive liberty, contributing to true self-realization. In this sense, nudging can be presented as an instrument that helps individuals take action to control their lives. Nudging can indeed serve as a tool for promoting positive liberty by empowering people to make more informed, self-beneficial decisions. This rationale is precisely what justifies its widespread use across such diverse decision-making contexts.

However, positive liberty focused on contributing to self-realization is also more prone to abuse based on the idea of liberating people to an idealized rational will. Ethical dilemmas arise wherein an external party deemed more capable defines what is in the best interest of an individual's well-being or what constitutes rational will. This risks compromising genuine autonomy and diminishing individuals' decision-making power by suggesting that an external agent knows best what aligns with their interests. The notion of delegating the understanding of what is in the individuals' best interest to governments or organizations while liberating the individual from self-mastering and responsibility for decision-making poses a significant risk to the exercise of liberty.

2.2 Freedom as Non-dominance

Freedom also includes another dimension, **non-dominance**, which refers to a state in which an individual is free from arbitrary domination by others, and no one has the power to potentially impose their will, even if that power is not actively exercised (Pettit 1996, 2012). Non-dominance emphasizes that true freedom requires the absence of potential domination, not just the absence of direct interference, which reflects the republican idea of freedom.

In the nudging context, non-dominance refers to the power dynamics inherent in the relationship established in nudging interventions. Nudging can be seen as an instrument of control (Grüne-Yanoff 2012; Pettit 2014; John 2018), where there is an unbalanced relationship between the nudger who designs the choice environment and dominates the influence that can or is potentially be exerted and the nudgee who is subtly dominated in a way that undermines her or his free will.

The concept of freedom as the absence of dominance has profound implications for the discussion of the ethics of nudging interventions. At its core, nudging presupposes that inherent imbalance between the **nudgees**, who are perceived as having limited decision-making capacity and need to be conducted in their path, and the **nudgers**, who intervene to signal the path to follow. Through various mechanisms, the nudgers guide those deemed less motivated, less rational, and more prone to behavioral biases toward making supposed better decisions. In this framework, the nudgers—governments, public institutions, or private companies—are positioned as authorities guiding the decision-making of the nudgees.

The critique stems from the power dynamics between the dominant (those who nudge) and the dominated (those being nudged), raising concerns about the imbalance of influence and control in shaping personal decisions. As Pettit (2011, p. 693) states, non-dominance implies "*that each option be accessible and that no one have the power to block access; the doors should be open, and there should be no powerful doorkeepers*". A profound ethical question arises at the level of the impact generated on society, where individual freedoms and pluralism are seen as central to preserving human dignity in the face of centralized powers and monochromatic views of life (Berlin 1969). If this balance of power is significantly skewed, concerns about how **doorkeepers**—also designated as **gatekeepers** to express the disproportionated power they wield—exercise their authority require greater attention and closer examination.

The non-dominance analysis can be analyzed separately in terms of the relationship of powers generated when used in public policy or by private institutions that operate on a market basis. From a **public policy perspective**, nudging is often justified as a means of promoting citizens' self-interest and the broader social good. In democratic contexts, nudges can serve as a gentle, non-coercive tool to guide citizens toward decisions that enhance societal welfare, gaining greater acceptance than more coercive or restrictive policies that are implemented in parallel. By subtly encouraging better choices, nudging aims to provide a better balance between guiding behavior and preserving individual freedom. Nudging can be seen as the lightest form of influence, where public powers preserve the individual's autonomy and do not exercise dominance as extensively because many of these policies could also be implemented coercively through other legislative and regulatory mechanisms at their disposal. Nudging interventions in public policy aim to gain citizens' cooperation towards solutions in their own best interest, supposedly as judged by them themselves, while also improving overall societal outcomes. And, these interventions offer an alternative to legislative solutions, which impose much stronger restrictions on freedom of choice. From a power and dominance perspective, public authorities hold significantly greater power, as they can legislate and coercively enforce policies. However, by opting for nudging strategies instead of direct imposition, they partially abdicate their position of dominance and grant individuals more autonomy in their decision-making. Additionally, as Schmidt and Engelen (2020) point out, there is no reason for public nudging policies to be more covert—and by that means affecting individual autonomy—than traditional laws and regulations, which allow nudging to remain democratically controlled.

From the **perspective of the private sector**, nudging also tends to be aligned with the institution's self-interest. Market nudges are more likely to bypass individuals' ability to deliberate and maintain control since they often lack the ethical principles governing public nudges, like softness and alignment to individual and public interest, and are frequently deployed in overwhelming numbers, making them harder to resist (Ivanković and Engelen 2024).

In sectors such as banking and fintech, for example, nudging is used to guide customers toward making financial decisions that are framed as beneficial for them, such as saving more or investing wisely. At the same time, these decisions also advance the institution's goals, such as increasing profitability or customer retention. This overlap can blur the line between what serves the customer and what benefits the organization, making it difficult to fully disentangle the two interests. While nudging may appear to promote better decision-making, the potential conflict arises when it becomes unclear whether the nudge primarily serves the customer's best interests or subtly pushes them towards actions that favor the corporation. Furthermore, in the private sector, well-intentioned nudges are often difficult to distinguish from well-structured marketing campaigns that claim to serve the customer's interest but are primarily designed to trigger actions that are in the organization's interest. Thus, from the point of view of dominance, there are two fundamental components that must be considered: the dynamics of the power relationship between the nudger and nudge, as well as the way that power is effectively exercised.

2.3 "As Judged by Themselves" and Heterogeneity of Preferences

The aim of nudging is to guide people towards a particular choice, often perceived as the best option. It is commonly referred to as the "***better off, as judged by themselves***" (Sunstein 2015). Promoting a choice as better or optimal assumes a hierarchy of options organized according to the utility that can be derived from each option. It also assumes that this hierarchy is stable across individuals or groups of individuals who are subject to the intervention. The nudging intervention is designed to point out the best option. Therefore, the purpose of nudging is to help people make more rational choices from which they can derive higher utility, choices that they supposedly would make if they had complete information, sufficient time to analyze it, and adequate cognitive resources to process it. Nevertheless, the supposition that an external observer is aware of the individual's genuine preferences overlooks the existence of diverse preferences. The nudging architect will typically assume that the decision-making process is flawed and that the individual should be guided toward more rational choices. For example, to encourage individuals to save more or to make healthier choices. However, some individuals may already engage in excessive saving or make choices that prioritize health to an excessive—sometimes unhealthy—degree. In such circumstances, a universal approach may prove inadequate.

Sunstein (2018) argues that when individuals' **antecedent preferences** are known in advance and they lack the self-control to implement them, nudges that align with these preferences can be ethically appropriate. The most complex case occurs when antecedent preferences cannot be observed or are unstable, making it challenging to operationalize a presumed optimal solution. Digital nudging and hypernudging, powered by AI, can collect specific information about individuals' antecedent preferences and create custom interventions tailored to each person accordingly. Hypernudging, in particular, allows for dynamic and continuous data collection, enabling real-time adjustments to tailor nudges to each individual in an ongoing and adaptative manner. However, even with knowledge of antecedent preferences and in fully individualized versions of hypernudging, the assumption that past preferences should dictate future choices presents a limited view of preferences, potentially narrowing the range of future options. Just because an antecedent preference indicates that the chosen flavor of ice cream is vanilla does not mean the next choice should also be vanilla. In many cases, changing preferences—whether for a reason or simply at random—is part of human nature and should be respected, and even encouraged, to foster a more diverse and lively world. While a nudge algorithm could detect the antecedent preference and suggest differently if instructed to do so, the question remains: what governs whether the algorithm should maintain vanilla as the next suggested choice or propose a different flavor? This raises broader concerns about who determines the criteria for when to reinforce consistency and when to promote variety, touching on deeper ethical questions about autonomy and control in personalized nudging systems.

When antecedent preferences are unknown, the assumption of guiding towards more rational choices and defining what is most rational for the individual is more challenging but reasonable when directed to values that tend to be generally accepted, such as health, wealth, and well-being. As nudging has the inherent possibility of opting out without significant cost or effort, keeping the other options available, it can also be ethically acceptable even without antecedent preferences, although transparency and acceptability take on a greater dimension in these circumstances. When antecedent preferences are unknown, the assumption of guiding individuals toward more rational choices—and defining what is most rational for them—becomes more challenging but can still be considered reasonable when it aligns with broadly accepted values such as health, wealth, and well-being. Since nudging inherently allows individuals to opt-out without significant cost or effort while keeping other choices available, it can be ethically acceptable even in the absence of antecedent preferences. However, in such cases, transparency and acceptability take on greater importance to ensure ethical integrity and trust.

In fact, through digital means, the observation of antecedent behaviors and preferences has become significantly more accessible. Even when such data is unavailable for a particular user, profiling can still be conducted based on alternative relevant attributes, allowing for the user categorization and extrapolation of her preferences. This can bring digital nudges closer to individuals' preferences, positioning them as better off, as judged by themselves. However, it may also make them susceptible to subtle, insidious, and manipulative influences.

3 Artificial Intelligence Driven Digital Nudging

3.1 Generative AI and Digital Nudging

Generative AI is changing how digital nudging operates by advancing the level of personalization used in nudging techniques. Automated assistants, like ChatGPT or AI copilots, offer users helpful suggestions and can improve decision-making competence by providing relevant support. However, if not thoughtfully designed and used, they may also risk diminishing users' decision-making competency by fostering reliance on automation over critical thinking.

In digital interaction, the observation of antecedent behaviors and preferences is readily available, along with other individual data that enables the profiling of users for more enriching, highly personalized experiences that can exert influence with precision. Digital platforms make use of visual cues, default settings, alerts, and reminders to encourage targeted behaviors. They also make use of user experience design principles – such as usability, simplicity, consistency, feedback, and emotional design—to make nudges and hypernudges more engaging and effective. Hypernudges based on the processing of large amounts of individual data generate hyper-personalized nudges and produce continually ongoing interactions, where each interaction is also new information to progress the nudging process further. This iterative and adaptive process not only customizes nudges in real-time but ensures that they are increasingly personalized to the targeted individual, improving the effectiveness of each subsequent interaction because every new cue, default, or recommendation becomes more accurate.

Digital environments become increasingly engaging and persuasive. The immersive choice environments are designed to improve user experience and guide decision-making. They make use of tools such as gamified choice designs, AI-based personal assistants, recommender systems and social (ro)bots that leverage on emotional cognition.

Co-pilots and virtual assistants powered by generative AI facilitate time savings and provide immediate responses across diverse topics. This may contribute either to the increase or decrease of individuals' agency:

- **Arguments for increasing agency**

  Virtual personal assistants may produce highly personalized hypernudges such as recommendations, helping individuals to make better informed decisions aligned with their goals and preferences. AI assistants can process a vast amount of data, allowing for better exploration of options and helping to produce more rational choices that are aligned with individuals' true intentions. Additionally, having been assisted in simpler or more recurrent decisions, where goals and preferences are established, individuals free their limited cognitive resources to focus on more important and complex decisions that require exploration and critical thinking is required.

- **Arguments for decreasing agency**

  Individuals feel supported by AI hypernudges and are more likely to accept suggestions as they seek to minimize the effort required in investigating options and examining various decisions. Frequently, these responses are perceived as accurate and are treated as reliable, which can diminish individuals' critical thinking skills. Hence, individuals may experience a sense of diminished responsibility for their choices, leading to a passive acceptance of cues and recommendations. This reliance on AI-generated information may lead to discounting their own judgments in favor of what is suggested and, consequently, reducing independent critical thinking.

These views are not mutually exclusive. They often overlap and intersect depending on the circumstances.

3.2 Nudging Processes and Outcomes

In nudging evaluation, there are two main dimensions to consider to reflectively evaluate their ethical implications: the process by which the nudges are implemented and the results they produce (Engelen 2019). An ethical nudge should satisfy both means and ends, which implies following an ethical process and producing an ethical outcome.

Process-based ethical nudges and **outcome-based ethical nudges** can coexist, yet it is also possible for one to be present while the other fails. In ideal scenarios, both the process and outcome are ethically aligned: the nudge is transparent and respectful of autonomy (process-based), and it leads to a positive individual and societal outcome (outcome-based). However, sometimes, a nudge may produce a positive outcome, for instance, steering a healthier choice, but still using questionable processes, such as for instance, a manipulative framing to generate those target healthier choices. Conversely, a nudge might be implemented ethically but fail to produce meaningful positive outcomes.

There is a tendency to prioritize outcomes, especially when these outcomes align with widely accepted social values like better health or wealth. Nudging for these universal goals tends to get greater acceptance, even in cases where the nudging process is less careful. The result is considered so relevant that it somehow justifies neglecting ethical features in the process in order to improve effectiveness. However, even when nudges promote individuals' overall well-being, they can still be possibly wrong if they are at odds with agency and/or autonomy.

In standardized nudges—where everyone receives the same prompt—the intended outcome and actual outcome may vary significantly across individuals. For instance, reducing food waste is seen as a valid intended outcome, and nudging for that purpose can be expected to be generally accepted. Reducing plate size might look like a good strategy to nudge reducing food waste. This nudging process can be implemented transparently and in full respect for autonomy by allowing to take additional helpings as individuals wish. However, suppose the reduction of the plate size is considered too large for some of those being intervened and that they choose to refill their plates. In these cases, the outcomes may become contrary to the intended goal, potentially leading to an overall increase in food waste rather than a decrease.

Hypernudges surpass this drawback by continuously tracking intended versus actual outcomes and personalizing interventions accordingly. Hypernudges are able to continuously adapt the incentives' design to better align with desired goals, improving the effectiveness of the nudging strategy. The ability to adapt to specific contexts and individual behaviors allows to tailor continuously updated interventions that maximize individual effectiveness. By analyzing user interactions and preferences, hypernudges can fine-tune, ensuring that nudges resonate with each user's unique circumstances and preferences. Thus, hypernudging may look like a panacea for processes' personal adjustments. However, continuous hypernudging based on individuals' observation and ongoing adjustments to targeted outcomes may also lead to a continuous insistence that degrades the nudging effectiveness and generates resistance due to exhaustion.

Furthermore, in terms of process, ethical hypernudging cannot be based on black boxes, wherein the influencing process remains unknown. Explainable AI (XAI) becomes necessary for ethical nudging processes. XAI enhances transparency by providing identification of how AI systems operate in nudging processes (for example, by describing the process to produce a recommendation). When individuals understand how their data is being used and how judgments are made on their behalf, they are better equipped to engage with the nudging initiatives as active agents.

In evaluating nudging effectiveness, it is also necessary to consider eventual hidden costs that refer to unforeseen side effects of nudging (Damgaard and Gravert 2018). Often, the effectiveness of nudges is assessed based uniquely on targeted outcomes, operating under the assumption that all other variables remain constant. This approach frequently overlooks the potential effects on adjacent variables, which may be affected unexpectedly and are often overlooked in ex-post evaluations. In the previous example, an unexpected effect could arise when individuals, in response to the nudging intervention with smaller plates, compensate by consuming additional snacks between meals. These snacks may be unhealthy—an unforeseen consequence—and could contribute to overall waste, which is an impact disregarded in the assessment of the targeted outcome. Such a narrow focus can lead to misunderstandings about the broader implications of nudging strategies, highlighting the need for a comprehensive assessment that accounts for all interrelated factors.

The interventions aimed at shaping behavior must consider these two dimensions in their ethical evaluation: the ethicality of the process and the ethicality of the outcomes, maintaining a broad perspective on processes and outcomes, namely, their collateral and long-term effects.

The long-term assessment is key, especially in cases of recurrent hypernudging based on generative AI with ongoing adjustment processes, where the long-term effects may be substantial and unexpected. A central ethical concern is evaluating the ongoing impact on agency and autonomy: whether the nudging intervention is benefiting self-determination, helping individuals to act as more informed and rational agents, or, conversely, turning them into passive receivers who simply accept suggestions as decisions. The response will define the level of acceptance that individuals and society have toward nudges and, more widely, the kind of society that is collectively created.

3.3 System 1 and System 2 Nudges

Nudges operate at different levels depending on the type of behavior they are addressing and how they intend to intervene. Kahneman's (2011) concepts of system 1 and system 2 thinking are determinants in defining the type of nudge intervention being implemented. **System 1 thinking** is fast, automatic, intuitive, and generates reflexive choices. In this system, nudges aim to display automatic behavior, such as habits and unconscious responses, by conducting decisions through subtle cues without requiring deliberate reflection or deep cognitive effort. Many nudges are designed to work in system 1, which generates ethical considerations as they are "invisible" nudges that target unconscious, reflexive responses. For instance, when signing up for a service, the automatic opt-in option for newsletters capitalizes on non-demanding cognitive effort. Differently, **system 2 thinking** is slow, deliberative, and logical. Nudges designed to engage system 2 encourage thoughtful consideration, such as informational nudges that aim to generate more reflected thinking. System 2 nudges are generally viewed as more transparent and ethically acceptable, as they promote conscious decision-making and respect individual autonomy. Instead of subtly pointing out the choice, system 2 nudges provide tools for critical thinking and informed decision-making. For instance, financial calculators or comparison tools in digital environments encourage in-depth analysis and logical reasoning, which is highly effective in complex decision-making. Therefore, from an ethical perspective, type 2 nudges tend to be preferred to type 1 nudges as they interfere to promote attention to the choice process and not so much to subtly point out the choice as system 1 nudges do. Hence, nudging designs, especially digital nudging and hypernudging, should consider the targeted thinking system for creating compelling and ethical nudges. Nudges can work with automatic processes or engage reflective capabilities, depending on the context, desired outcome, and ethical considerations.

The level of autonomy experienced may also differ depending on the type of nudge used, as they do not all operate in the same way nor have the same impact on decision-making processes. Nudges can be classified as targeting system 1 or system 2 thinking (Kahneman 2011) depending on whether they are heuristics-based or informational-based (Leal and Oliveira 2021). **System 1 nudges** target a more automatic and less conscious level of decision-making, utilizing heuristics and cognitive shortcuts. By influencing decisions in a more subtle way, often without individuals realizing they are being "nudged", these nudges can be seen as more intrusive. Their effect on the less reflective areas of the brain raises ethical concerns, as they can be perceived as hidden interventions that exploit cognitive vulnerabilities without allowing space for critical reflection. In contrast, **system 2 nudges** tend to provide additional information or structure the decision environment in a way that supports cognitive processes, supporting more profound and deliberate reflection. These nudges produce more informed and conscious choices, thus respecting and reinforcing individual autonomy. By improving the quality of the decision-making process, such nudges support autonomy and enable more informed and deliberate choices, ultimately reinforcing positive liberty. The use of system 1 or system 2 nudges signals the role that the choice architect intends people to play when developing personalized nudges and hypernudges.

3.4 Ethical Concerns in Nudging Personalization

The primary ethical concern in AI-driven digital nudging lies in balancing the benefits of personalization with the preservation of autonomy. The in-depth personalization risks crossing the line to become dissimulated manipulation. This risk is particularly pressing when AI-driven assistants may assume decisive control over decision-making. There is a risk of over-personalization, where nudging may become intrusive or manipulative, undermining user autonomy. Attention is also needed to the ethical implications of using personal data to influence decisions subtly, questioning when nudging crosses the line into manipulation.

Many personalized nudges aim to stimulate fast, automatic System 1 thinking, often producing **mindless reactions**. These personalized nudges rely on cues, defaults, and recommendations, generating reactive behaviors with minimal deliberation. In these circumstances, the risk of undesired and misleading influence becomes a significant concern. However, as nudges, they should also be easy to avoid. On the other hand, nudges can also be constructed to activate System 2 thinking, supporting **mindful reflection**. These are mainly informational nudges that place the individual at the center of the decision process, inviting them to ponder deeply. Through digital nudges and hypernudges, information can be presented to meet individual needs, thereby providing relevant insights for thoughtful consideration, representing an increase in autonomy in the sense of self-governing to achieve self-realization.

System 2 nudges represent a more ethical approach to nudging. However, even informative system 2 nudges can become problematic. In an increasingly AI-driven world, dependence on AI personal assistants is expected to grow. As individuals become used to receiving highly precise, tailored information, they may also become more dependent on these tools, increasingly willing to receive not only curated information but the decision itself. This growing dependency reflects a shift where individuals might prefer solutions provided directly rather than actively engaging in critical evaluation, potentially representing an autonomy reduction.

Additionally, in the process of informing, the presentation and formatting of information are often strategically leveraged. Even when truthful information is used, selectively structuring it to meet specific objectives can lead to an emphasis on more convenient truths. In parallel, other equally relevant details are downplayed or omitted. This might result in the **misuse or abuse of truth** for targeted purposes, where formatted truths or partial truths are presented.

There is a fine line between helping and exploiting decision-making processes. The existence of a boundary does not imply a straightforward, well-defined border where the ethical version of each intervention is clear. Navigating the complexities of behavioral science interventions requires a close ongoing examination of intentions, processes, and outcomes.

Zuboff (2019) noted that the data-driven nature of these technologies can lead to a form of surveillance capitalism, where individual choices are manipulated based on predictive analytics, compromising true agency. Hypernudging through self-tracking technologies may undermine individual autonomy by infringing on both informational and decisional privacy (Lanzing 2019). This erosion of privacy may limit the ability to make independent choices, as individual data is continuously monitored and nudged toward certain behaviors.

4 Legal Framework for Digital Nudging in the European Union

4.1 The European Union's Role in Defining Standards for Digital and AI Practices

Ethical concerns about the rapid evolution of digital platforms, automated assistants using natural language processing, and the proliferation of AI-augmented systems have urged the advance of a robust legal framework. The European Union (EU) has been taking a leading role in translating these concerns into legislative action across various regulations, responding as new technological realities develop. Although no specific regulation addresses nudging directly, several EU legislative initiatives address the issues involved in nudging practices, providing users' protection and establishing guidelines for fair digital interactions (Leal and Oliveira 2024).

The following are key pieces of legislation translating various ethical concerns into legal standards, underscoring the EU's commitment to protecting individual rights and promoting responsible digital practices addressing digital age risks: (1) General Data Protection Regulation; (2) Digital Markets Act; (3) Digital Services Act; and (4) Artificial Intelligence Act.

4.2 General Data Protection Regulation

The European Union's first legislative initiative on data protection dates back to 1995, with Directive 95/46/EC (European Parliament 1995). This directive established rules for personal data protection, aiming to safeguard privacy and improve data protection rights while facilitating the free movement of personal data across member states. It emphasizes that data must be processed fairly and transparently, collected for specific purposes, and limited to what is necessary for those purposes.

The EU Charter of Fundamental Rights (European Union 2012) establishes the protection of personal data as a fundamental right under Article 8, reinforcing the importance of privacy and the safeguarding of individuals' personal information. It reinforces the notion that individuals should have control over their personal data as a fundamental freedom, reflecting growing societal awareness of privacy issues in the digital age.

The evolution of expectations regarding personal data handling and privacy, along with technological advancements in the rise of digital environments, has driven the enhancement of individuals' rights over their personal data, as encapsulated in the **General Data Protection Regulation (GDPR)**—Regulation (EU) 2016/679 (European Union 2016). This regulation aims to adapt the EU to the digital age, focusing on the protection of natural persons with regard to the processing of personal data and on the free movement of such data across the European Union and repealing Directive 95/46/EC. The GDPR, along with guidelines from the European Data Protection Board (EDPB), **mandates** transparency, user consent, and rights over one's data, including access, rectification, and erasure.

GDPR restrictions ensure that digital nudging reliant on personal data only occurs with user consent. The guidelines emphasize that consent should be freely given and specific, reinforcing user control and limiting manipulative nudging based on personal data, particularly concerning the use of data for unspecified purposes or cross-referencing between organizations. This legislation restraints aggressive digital nudging and hypernudging, especially when it comes to using data for unspecified purposes or exchanging individual data to build more robust databases through which they can target prospective individuals more precisely.

4.3 Digital Markets Act

The Regulation (EU) 2022/1925 on Contestable and Fair Markets in the Digital Sector, known as the **Digital Markets Act—DMA** (European Union 2022a), regulates digital markets and large online platforms—designated as gatekeepers—aiming to ensure fair competition. This gatekeepers designation alludes to a concept similar to doorkeepers, as discussed by Pettit (2011) in the exploration of the notion of non-domination (as discussed in Sect. 2.2). Pettit argues that for a state of non-domination to be achieved, it is fundamental not only that the doors remain open but also that there are no powerful doorkeepers who might exert undue influence or control over access. In light of the immense power wielded by these platforms, the discussion now shifts to the term gatekeepers, highlighting their substantial power in shaping access and decision-making within digital spaces.

In September 2023, under the Digital Markets Act, the European Commission designated six **gatekeepers**—Alphabet, Amazon, Apple, ByteDance, Meta, and Microsoft (European Commission 2023). These tech giants, with their vast resources and significant influence over individual data processing, including advancements in cutting-edge AI technologies they develop, wield unprecedented control over users' behavior. The asymmetry of power has never been greater, and society is currently navigating towards a new paradigm. This shift reflects significant changes in the dynamics of power in digital technology due to not only a profound change in the capabilities for processing vast amounts of data but also in the concentration of this power within a limited number of entities.

This act limits how digital platforms—the gatekeeperscan use their dominant positions to influence user choices unfairly. Enforcing transparency and accountability reduces the potential for manipulative nudging by promoting fairer competition and consumer autonomy.

4.4 Digital Services Act

The EU Regulation 2022/2065 on a Single Market for Digital Services, known as the **Digital Services Act—DSA** (European Union 2022b), establishes new responsibilities for online platforms, including rules for content moderation, transparency, and user rights, creating a safer and more transparent online environment.

The DSA aims to rebalance the roles of users, digital platforms, and public authorities according to values that place the citizens at the center. It provides better protection of European fundamental rights and stronger protection from illegal content. Overall, it fosters a digital environment that respects both individual rights and collective democratic values (European Commission 2024).

The DSA requires digital platforms to disclose information about how content algorithms function. This transparency is key for digital nudging as it helps users understand and control the nudges they are exposed to, reducing the risk of manipulation and enhancing informed decision-making.

While the DMA focuses on market competition and a level playing field, the DSA emphasizes user protection and transparency. Despite their differences, both laws recognize the importance of transparency in the practices of online platforms and aim to create a fair and competitive digital marketplace. Regarding implications for digital nudging, the DMA promotes fair competition, which may lead to more ethical nudging practices as companies seek to differentiate themselves through transparency and user trust. On the other hand, the DSA's focus on user protection may limit manipulative nudging techniques that exploit user vulnerabilities. Jointly, both frameworks aim to create a fair and competitive digital marketplace, potentially leading to more responsible and ethical approaches to nudging in digital environments.

4.5 *Artificial Intelligence Act*

The EU Regulation (EU) 2024/1689 on laying down harmonized rules on artificial intelligence, known as the **Artificial Intelligence Act—AI Act** (European Union 2024), is the first comprehensive AI law. The AI Act regulates the use of artificial intelligence, especially in high-risk systems that affect fundamental rights, categorizing systems based on risk levels and defining standards for each category. These regulations are designed to prioritize the safety and rights of individuals while enabling responsible AI innovation within the EU.

AI systems classified as Unacceptable Risk AI are considered potentially harmful to people and are therefore banned. These systems include:

(1) *Cognitive Behavioral Manipulation* of people or specific vulnerable groups, such as tools that encourage risky behaviors in children;
(2 *Social Scoring* for classifying people based on behavior, socio-economic status, or other personal attributes, potentially leading to biased treatment; and
(3) *Biometric Identification and Categorization*, including technologies like facial recognition (European Parliament 2024).

Moreover, all high-risk AI systems will undergo a thorough assessment before market entry and continuously throughout their lifecycle. Even though they are not categorized as high-risk, generative AI systems must comply with transparency and copyright requirements, namely:

(1) *Disclosure* to ensure users are informed when content is AI-generated;
(2) *Content Safeguards* mechanisms to prevent the generation of illegal content; and
(3) *Copyright compliance* by publishing summaries of copyrighted material used during training (European Parliament 2024).

For nudging applications driven by high-risk AI, the AI Act enforces rigorous requirements to prevent harm to user autonomy and avoid exploiting cognitive biases. This directly impacts some hypernudging practices based on profiling, ensuring high ethical standards and preventing abusive digital nudging.

Ensuring compliance with all legal obligations requires *constant monitoring*, supported by robust enforcement measures. Continuous oversight is an effective deterrent, pressing organizations to follow established standards consistently.

4.6 Worldwide Impact

This EU legal framework has intrinsic value in protecting individuals within the European Union and addresses many ethical concerns about nudging. Collectively, these regulations protect users' rights, privacy, and autonomy, creating a structured approach to digital markets in the EU.

Moreover, this legal framework has the potential to extend its **impact beyond the EU's geographical boundaries** in two ways. First, it applies to all providers, whether based within or outside the EU, as long as their services are directed to users within the EU. This requirement may pressure these companies, designated the gatekeepers and doorkeepers, to adopt similar practices even outside the EU, avoiding user discrimination based on geography. Obviously, large platforms may have different policies inside and outside the EU. However, that approach highlights that less regulated regions are at a disadvantage, with lower protection against certain potentially unethical practices. Therefore, platforms that aim to disclose better practices will likely adopt them proactively.

Second, it acts as a guide to fundamental safeguards, potentially serving as a legislative guide for other geographic jurisdictions that choose to regulate along these lines. Currently, various other countries have their own regulatory initiatives addressing these same concerns. The EU's comprehensive legal framework can serve as a benchmark, stimulating the dissemination of these regulatory concerns and their transposition into the legal frameworks of other countries.

5 Principles Guiding the Ethics of Nudging

5.1 Dimensions of Ethical Nudges

The ethical foundations of nudging are key to its acceptance by society and its effectiveness in guiding individuals toward better decision-making. The evaluation of whether a nudge is ethical or unethical hinges on a set of fundamental values within decision-making processes that ensure respect for **individual freedom of choice and decision-making autonomy**. At its core, and in a strict definition of nudging, the aim is to encourage better decisions without overly interfering and always allowing the easy option to choose otherwise. As such, the ethical boundaries of nudging should be well-defined, striving ultimately to serve the individual's best interest. This inherent alignment with the individual's best interest and welfare should ensure that nudging is good for individuals and society, especially compared to other regulatory and commercial policies. Nonetheless, the discussion surrounding fundamental guarantees of freedom of choice aims to ensure that nudging—both in conceptual terms and practical implementation—effectively serves people in the best possible way.

Public debate and scrutiny are needed to balance the intended benefits of nudging and the actual outcomes, ensuring no misalignment, particularly in the long run. This ethical dialogue centers on the basic rights of liberty, examining whether nudging interferes with these rights and how it ultimately shapes society.

As previously mentioned, the ethical examination can focus on either the process of influence or the outcome. Compared to alternative methods of shaping decisions at the process level, nudging is a less intrusive approach at the process level. At the outcome level, it can more effectively maximize expected utility. Consequently, it is reasonable to expect greater societal acceptance compared to more intrusive methods of influence and regulation. This is, at least, the core argument made by proponents of nudging.

An ethical evaluation of a nudge may imply several considerations (for instance, Leal et al. 2022). That does not imply that a checklist properly classifies a nudge as ethical or unethical. However, the following are some features that should be taken into consideration in a nudge's ethical evaluation.

5.2 Beneficial Intention

Ethical nudges are designed with the genuine intention to benefit individuals and society, helping to make choices aligned with their long-term well-being. For example, encouraging savings for retirement. The primary aim is the well-being of the individual, pursued in a way that neither compromises the welfare of others nor undermines societal interests. The individual's benefit is pursued in harmony with the common good, ensuring both personal and collective well-being are aligned.

Beneficial nudges that align with individuals' own interests and manipulative nudges that primarily serve nudgers' commercial interests often coexist and disentangling them is not straightforward. In many cases, the same nudge may serve dual purposes. For instance, on online platforms like Amazon, users are shown additional information under the labels of "*products related to this item*", "*customers that viewed this also viewed*", "*customers who bought this item also bought*", and "*frequently bought together*", as well as other customers and experts reviews. These recommendations may provide helpful suggestions and facilitate navigation across options, but they also have obvious commercial motivations. Some platforms take this further, suggesting purchasing bundles of related items or replacing the shopping cart with one that includes complementary items, ready for purchase at the click of a button. Reminders for items abandoned in the shopping cart or notifications suggesting it is time to reorder a product add to this hypernudging dynamic, combining convenience with nudges that benefit the platform. This dual purpose exemplifies the ethical complexities in digital nudging, where helpful information and persuasive tactics often overlap.

5.3 *Respect for Autonomy*

Ethical nudges respect individual autonomy by preserving freedom of choice, which means that all options remain available and it is easy to choose any option. For example, easily opting-out of a default option. This ensures that while individuals are gently guided toward a beneficial choice, their right to make an independent decision remains fully intact, preserving the essence of self-determination.

In an increasingly technology-based world, individuals also need to embrace their role in decision-making and recognize that AI assistants and recommendation systems are intended to support, not replace human judgment. Responsibility ultimately rests with the individual, who should maintain decision authority and disregard defaults or recommendations when considered unappropriated. The existence of AI-supporting systems should reinforce the relevance of each person's active engagement as the primary decision-maker rather than over-relying on AI. Additionally, it is critical to disentangle nudging from more intrusive forms of influence or manipulation, which differ from the strict definition of nudging and do not foster genuine choice or autonomy.

5.4 *Transparency*

Ethical nudges are transparent, making individuals aware of the influence that is being exerted. They should acknowledge either their intentions or mechanisms of action to attain better outcomes. For example, when selecting a default option, it should clearly communicate why it is considered the most beneficial while highlighting that individuals are free to choose any other alternative. This level of openness respects personal autonomy and strengthens trust in the process, ensuring that the nudge is seen as a helpful guide rather than covert manipulation. Moreover, research suggests that transparency in nudging does not diminish its effectiveness (Bruns et al. 2018) while clearly strengthening ethical standards.

5.5 *Acceptance*

Ethical nudges seek acceptance of nudged participants and the general society. With this purpose are willing to disclose information about their processes and outcomes. For example, reports disclosing the nudges that are implemented and the results implemented for fostering retirement savings. By openly communicating their intentions, methods in place and outcomes attained to achieve, ethical nudges foster trust, ensuring that individuals and communities feel engaged and respected in decision-making. Thus, transparency enhances societal acceptance and strengthens the legitimacy of the nudge itself. This acceptance can be reinforced by societal watchfulness (Ivanković and Engelen 2019) where individuals actively monitor the use of nudges and have the capacity to accept or resist if they understand that nudges are undermining their power of choice.

5.6 *Informed Consent*

Ethical nudges aim for individuals to understand that they are being guided and willingly accept that guidance. Simply put, in any nudging intervention, there must be expressed and informed consent for influence to be exercised. Even subtle nudges may require informed consent regarding the acceptance of being nudged. It reinforces the quality of the relationship between the nudger and the nudgee and, therefore, may reinforce the nudging effectiveness as it responds to an expressed desire to be nudged.

The requirements for informed consent declarations are expected to become increasingly rigorous to ensure individuals fully understand how their data is used, particularly to prevent any misuse of private information for profiling and individual personalization. Thus, requiring users to check boxes indicating that they have read or agree with privacy policies—often referencing long and complex documents that users many times confirm without actually reading—will likely be replaced by other approaches, such as shorter, clearer declarations that foster a more active and informed consent process.

Informed consent must be reinforced even further when handling sensitive data, such as health or financial data, emphasizing the importance of safeguarding against any misuse, especially in cases where information is considered sensitive. For nudging based on sensitive data, it may even be necessary for users to demonstrate consent more actively, such as typing their declaration of consent into a designated text box. This approach aims to ensure that users genuinely understand how their data will be used and how nudging is going to be implemented, thereby minimizing potential misunderstandings and promoting truly informed consent.

5.7 *Respect for Vulnerable Groups*

Ethical nudges consider the vulnerability of certain groups and ensure that they are protected and supported in making informed decisions. They do not exploit any cognitive or affective vulnerabilities. For example, additional requirements in terms of consent must be provided to vulnerable older adults. By adjusting interventions to the unique needs of individuals, ethical nudging not only safeguards autonomy but also promotes equity.

Overall, compliance with these ethical principles ensures that nudges are used responsibly, safeguarding fundamental freedoms, self-determination, dignity, and protection from abusive practices, especially for vulnerable people.

6 Responsible Nudging: A New Agenda for Disclosure and Certification

In the current age of influence, where AI allows the processing of a vast amount of data for comprehensive individual profiling, it is natural that society is raising concerns about these persuasive forces. Within this context of varied influence strategies, nudging presents itself as a soft form of choice environment designed for behavioral change more aligned with individuals' best interests. Thus, ethical concerns should presumably focus on other stronger intervention methods, particularly those driven by commercial objectives that frequently do not align with the interests of the targeted individuals but with the goals of those who put them in action.

Nevertheless, as nudges often claim to be beneficial and non-intrusive (without restraining options or significantly changing incentives), they are frequently under more intense scrutiny to ensure that these claims are genuinely reflected in practice. The ethical discourse surrounding nudging discusses the fundamental values of freedom, agency, autonomy, and preferences, especially as it employs mechanisms similar to other forms of influence. Ensuring that nudging genuinely supports decision-making without eroding personal agency or autonomy is essential to maintaining it as a tool of empowerment rather than a subtle form of control.

This ethical examination is essential to ensure that nudging fulfills its promises, positioning nudging as a truly supportive tool in decision-making. To this end, we argue that nudging interventions should always be explicitly identified as such, describing how the guiding ethical principles are addressed. **Responsible nudging** interventions must provide a framework that ensures adherence to nudging's ethical principles. Such a structured ethical framework legitimizes nudging practices, allows for watchfulness and fosters public trust by anchoring these interventions in high standards of accountability and ethics.

To ensure that nudging employs the best practices and succeeds in intended outcomes, we propose the label of responsible nudging that could be associated with each intervention that complies with comprehensive ethical requirements. To do so, two main features should be observed, especially in digital nudging and when AI-generated hypernudges are in use:

- **Responsible nudging disclosure**

  We argue that, beyond meeting rigorous legal standards, responsible nudging must demonstrate compliance with ethical requirements that should be thoroughly documented and auditable by external organizations. Each nudging intervention should be thoroughly disclosed in order to provide clear information about the nudging practices in use to current users, potential users, supervisory entities, and society. Besides, organizations should disclose a comprehensive summary of nudging their strategies in an Organization Nudging Report that details its aims and means. This report should describe the positive and negative impacts of nudging interventions like sustainability disclosure in Corporate Sustainability Reports. This Responsible nudging reporting would also be an instrument for helping organizations assess risks and opportunities and drive nudging strategies.

- **Responsible nudging certification**

  We argue that having independent entities accrediting and certifying nudging practices in terms of ethical principles would increase ethical standards. It would reinforce transparency, ensuring that interventions are truly aligned with ethical guidelines and societal expectations. These external entities would serve as guarantors for the responsible use of nudges, ensuring they contribute to individual and societal well-being without compromising ethical values such as self-determination and dignity. At a time when concerns about the power exercised by gatekeepers are prevalent, presenting themselves as accredited entities whose practices undergo verification and achieve responsible nudging certification would be a step toward rebalancing power dynamics and safeguarding citizens. Obtaining these responsible nudging labels would reinforce accountability and help establish public trust in the ethical use of nudges.

This agenda is closely aligned with the ongoing evolution of AI technology in digital nudging interventions. It requires regular updates in disclosure practices and certification processes to ensure that ethical standards remain relevant and effective in light of technological advancements.

7 Final Remark: The Ethical Challenges in Digital Nudging

In an age of influence and manipulation, nudging represents a comparatively minor concern when contrasted with other forms of influence that do not prioritize the individual's best interests—neither in the process nor in the outcome. Unlike these more insidious practices, nudging typically encourages behavioral changes to enhance individuals' well-being, as judged by themselves.

Regulatory efforts across various areas of behavioral influence will ensure that nudging considers a wide range of ethical concerns, many of which are interconnected with legal dimensions. Nevertheless, the ethical debate surrounding nudges—particularly digital nudges and hypernudging—will continue to be a pressing and evolving issue, with challenges that are intrinsically linked to the ongoing evolution of technology. The gatekeepers should look for going beyond mere legal compliance, demonstrating a genuine commitment to ethical considerations and transparency regarding their nudging practices. We argue that responsible nudging labels should serve as a guarantee of ethical standards in the implementation of nudges, assuring the public that such interventions follow principles of ethical nudging. Transparent disclosure and the certification of responsible nudging are central to distinguishing it from other forms of influence and earning the trust of users and society.

Acknowledgements This chapter was supported by grant no. 2022.07029.PTDC, supported by the Foundation for Science and Technology (FCT).

This chapter was edited with the assistance of Grammarly, an AI-powered writing tool, to enhance correctness and clarity.

Abbreviations

AI	Artificial Intelligence
DMA	Digital Markets Act
DSA	Digital Services Act
EDPB	European Data Protection Board
EU	European Union
GDPR	General Data Protection Regulation
XAI	Explainable Artificial Intelligence

References

Berlin, I. (1969). Two Concepts of Liberty. In Four Essays on Liberty (pp. 118–172). Oxford University Press.

Bovens, L. (2009). The ethics of nudge. In *Preference change: Approaches from philosophy, economics and psychology* (pp. 207–219). Dordrecht: Springer Netherlands.

Bruns, H., Kantorowicz-Reznichenko, E., Klement, K., Jonsson, M. L., & Rahali, B. (2018). Can nudges be transparent and yet effective?. *Journal of Economic Psychology*, *65*, 41–59.

Damgaard, M. T., & Gravert, C. (2018). The hidden costs of nudging: Experimental evidence from reminders in fundraising. *Journal of Public Economics*, *157*, 15–26.

Engelen, B. (2019). Nudging and rationality: What is there to worry?. *Rationality and Society*, *31*(2), 204–232.

European Commission (2023). *Digital Markets Act: Commission designates six gatekeepers*, 6 September 2023. Retrieved from https://ec.europa.eu/commission/presscorner/detail/en/ip_23_4328 (Accessed 15.10.2024).

European Commission (2024). *The Digital Services Act: Ensuring a safe and accountable online environment*. Retrieved from https://commission.europa.eu/strategy-and-policy/priorities-2019-2024/europe-fit-digital-age/digital-services-act_en (Accessed 15.10.2024).

European Parliament (2024). EU AI Act: first regulation on artificial intelligence, 18 July 2024. Retrieved from https://www.europarl.europa.eu/topics/en/article/20230601STO93804/eu-ai-act-first-regulation-on-artificial-intelligence#transparency-requirements-1 (Accessed 15.10.2024).

European Union (1995). *Personal Data*, Directive 95/46/EC of The European Parliament and of The Council of the European Union of 24 October 1995 on the protection of individuals with regard to the processing of personal data and on the free movement of such data. https://eur lex.europa.eu/legal-content/EN/TXT/PDF/?uri=CELEX:31995L0046.

European Union (2012). *Charter of Fundamental Rights of the European Union,* European Union, 26 October 2012 https://eur-lex.europa.eu/eli/treaty/char_2012/oj.

European Union (2016). *General Data Protection Regulation,* Regulation (EU) 2016/679 of The European Parliament and of The Council on the protection of natural persons with regard to the processing of personal data and on the free movement of such data, and repealing Directive 95/46/EC. https://eur-lex.europa.eu/legal-content/EN/TXT/PDF/?uri=CELEX:32016R0679.

European Union (2022a). *Digital Markets Act, Regulation (EU) 2022/1925 of the European Parliament and of the Council of 14 September 2022 on Contestable and Fair Markets in the Digital Sector.* https://eur-lex.europa.eu/legal-content/EN/TXT/?uri=uriserv:OJ.L_.2022.265.01.0001.01.ENG.

European Union (2022b). *Digital Services Act, Regulation (EU) 2022/2065 of the European Parliament and of the Council of 19 October 2022 on a Single Market for Digital Services.* https://eur-lex.europa.eu/eli/reg/2022/2065/oj.

European Union (2024). *Artificial Intelligence Act,* Regulation (EU) 2024/1689 of the European Parliament and of the Council of 13 June 2024 on laying down harmonised rules on artificial intelligence. https://eur-lex.europa.eu/eli/reg/2024/1689/oj.

Grüne-Yanoff, T. (2012). Old wine in new casks: Libertarian paternalism still violates liberal principles. *Social Choice and Welfare,* **38**(4), 635–645.

Hansen, P. G., & Jespersen, A. M. (2013). Nudge and the manipulation of choice: A framework for the responsible use of the nudge approach to behaviour change in public policy. European Journal of Risk Regulation, 4(3), 3–28.

Ivanković, V., & Engelen, B. (2019). Nudging, transparency, and watchfulness. *Social Theory and Practice,* 43–73.

Ivanković, V., & Engelen, B. (2024). Market nudges and autonomy. *Economics & Philosophy, 40*(1), 138–165.

John, P. (2018). *How far to nudge? Assessing behavioural public policy.* Cheltenham: Edward Elgar Publishing.

Kahneman, D. (2011). Thinking, fast and slow. *Farrar, Straus and Giroux.*

Lanzing, M. (2019). “Strongly recommended” revisiting decisional privacy to judge hypernudging in self-tracking technologies. *Philosophy & Technology, 32,* 549–568.

Leal, C. C., & Oliveira, B. (2021). Choice Architecture: Nudging for Sustainable Behavior. *Sustainable Management for Managers and Engineers* (p. 17). UK: John Wiley & Sons.

Leal, C. C., & Oliveira, B. (2024). Nudging financial behavior in the age of artificial intelligence. *Artificial Intelligence in Production Engineering and Management* (pp. 115–144). UK: Elsevier.

Leal, C. C., Branco-Illodo, I., Oliveira, B. M. N., & Esteban-Salvador, L. (2022). Nudging and Choice Architecture: Perspectives and Challenges. *Journal of Contemporary Administration, 26*(5 Special Issue), e-220098

Pettit, P. (1996). Freedom as antipower. *Ethics, 106*(3), 576–604.

Pettit, P. (2011). The instability of freedom as noninterference: the case of Isaiah Berlin. Ethics, 121(4), 693–716.

Pettit, P. (2012). On the people’s terms: a republican theory and model of democracy. Cambridge University Press.

Pettit, P. (2014). *Just freedom: A moral compass for a complex world.* New York: W. W. Norton & Company.

Schmidt, A. T., & Engelen, B. (2020). The ethics of nudging: An overview. *Philosophy compass, 15*(4), e12658.

Sunstein, C. R. (2014). Why nudge?: The politics of libertarian paternalism. Yale University Press.

Sunstein, C. R. (2015). The ethics of nudging. *Yale J. on Reg., 32,* 413.

Sunstein, C. R. (2016). The ethics of influence: Government in the age of behavioral science. Cambridge University Press.

Sunstein, C. R. (2018). “Better off, as judged by themselves”: a comment on evaluating nudges. *International Review of Economics, 65,* 1–8.

Thaler, R. H. (2008). Sunstein. C. (2008), Nudge: Improving decisions about health, wealth, and happiness, Yale University Press.
Thaler, R. H. (2018). Nudge, not sludge. *Science, 361*(6401), 431–431.
Yeung, K. (2019). 'Hypernudge': Big Data as a mode of regulation by design. In *The social power of algorithms* (pp. 118–136). Routledge.
Zuboff, S. (2019). The Age of Surveillance Capitalism: The Fight for a Human Future at the New Frontier of Power, edn. *PublicAffairs, New York.*

Corporations as Communicative Moral Agents

José Maria L. de Pedro

Abstract Recently, a small group of authors have begun to interpret the moral agency of corporations as a phenomenon that occurs through communication. With different approaches, these authors link the moral agency of firms to their ability to perceive and react to the normative expectations of other agents. These authors rightly emphasize the communicative nature of morality. However, they interpret it as a homogeneous phenomenon affecting only individuals and firms. Like these authors, I assume that morality always occurs through communication. In contrast, I offer a more complex view. In particular, I describe morality as a communicative phenomenon that occurs at three levels, determined by the object of communication and the functional capacities exercised by agents. At the expressive level, agents use expressions of esteem (such as praise, resentment or blame) to mutually evaluate each other's attitudes and behaviors, taking as a reference the normative structure that regulates their interactions. At the reflective level, they communicate about the normative structure that coordinates their expectations and behaviors when interacting. At the ethical level, they communicate with a general and abstract approach to some normative element, regardless of how that element materializes in the specific structure that regulates their interactions. Corresponding to those levels, I identify three modes of moral agency (expressive moral agency, reflective moral agency, and ethical agency). After presenting this framework, I argue that corporations often exercise these capacities in practice and should therefore be considered moral agents of each type.

Keywords Morality · Moral agency · Systems theory · Communication · Corporate moral agency

J. M. L. de Pedro (✉)
Universidad Villanueva, Madrid, Spain
e-mail: jmlopez@villanueva.edu

C. Machado (ed.), *Ethics in Management and Business*, Management and Industrial Engineering, https://doi.org/10.1007/978-3-031-82054-0_4

1 Introduction

In business ethics, it is frequent to divide authors into two groups regarding the issue of corporate moral agency (CMA). One group argues that corporations can be considered moral agents, as they exercise certain qualities such as self-control, normative autonomy, or rationality (e.g., Hess 2018; List and Pettit 2011). Another group claims that only individuals can fulfill the right conditions to be recognized as moral agents (e.g., Sepinwall 2017; Velasquez 2003). Despite their differences, both groups address the issue of moral agency by either ignoring communication or interpreting it as a linear process that agents use to convey messages without affecting the nature or meaning of their operations. In contrast to this simplistic view, a group of authors has recently started to see communication as a crucial phenomenon for understanding morality in business. Key examples of this turn are the proposals of *diachronic responsibility* (French 2017), the *conversable firm* (Pettit 2017), and the *second-person narrative* (Morrison et al. 2022). In all these proposals, the authors interpret corporations as discursively moral agents, as they can perceive other agents' expressions of what is acceptable or unacceptable, and react to those expressions.

In business, dynamics such as those described by the authors of this communicative turn often occur. For example, McDonald's has recently been harshly criticized by many individuals, activist groups and even institutional investors for its policies regarding the welfare of animals raised by its suppliers (Cain 2022). As a result of this dynamic, the firm has been able to redesign certain policies on the issue (Block 2024), recovering its adjustment to the normative demands expressed to the firm by other agents from its closest social context. The authors of the communicative turn have already recognized this type of dynamics in which a firm is the object of the reactive expressions with which other agents evaluate its behaviors. However, moral communication often affects firms in different ways. In many situations, for example, firms negotiate with other agents about the norms that apply to their interaction. This is the case, for example, of many large fashion companies (such as Nike, Inditex, H&M or Patagonia) that negotiate with investors, suppliers, governments, competitors or NGOs to design the normative structure that should guide their behaviors (e.g., Pehn 2023; Reuters 2024). In this and many other similar cases, corporations participate with other agents in communicative processes in which they discuss the normative structure (made up of norms, values, principles, examples…) that guides their mutual expectations and demands. Unfortunately, this type of dynamics has not yet been addressed by the authors of the communicative turn.

In this chapter, I adhere to the communicative turn to address the following questions: how does morality occur through communication? What functions must an entity exert to communicate morally? And, based on my answers to those questions, is it plausible to consider corporations as moral agents? I formulate the first and second questions in a very open-ended manner. This approach will allow me to make my reflection extensive to all types of entities, not just individuals and firms. In fact, my ideas on this issue will resonate—to the reader interested in moral theory—with many theoretical and empirical developments taking place in other fields of research, such as the study of morality in animals (e.g., Palao 2021; Wrage 2022) or the moral agency of artificial systems (e.g., Ladak 2023; Luccioni and Bengio 2020). I formulate the third question in a much more restrictive way, with the specific purpose of assessing whether firms can fulfill the necessary conditions to be considered moral agents.

To answer these questions, I construct a framework that connects communication and morality. Following Niklas Luhmann, I describe morality as a phenomenon that is essentially connected to communication. According to Luhmann, morality is a communicative resource for the social coordination of individuals. From this premise, I expand and correct some of Luhmann's ideas to construct a theoretical framework in which moral communication occurs at three functionally different but related levels. At the expressive level, agents of any ontology perceive and react to the expressions of esteem of others who evaluate their behaviors in reference to a normative structure (comprised of norms, values, principles, examples…) that guides their interactions. For instance, this occurs when a girl expresses her anger to a friend who is late for their scheduled meeting. At the reflective level, the agents communicate about some elements of the normative structure that rule their interactions. For example, this happens when two friends, after sitting in a café, directly discuss which behaviors should be deemed unacceptable in their relationship. At the ethical level, agents communicate with a general and abstract focus on any normative element, regardless of how it materializes in the normative structure that regulates their interactions. This would be the case, for instance, when a group of friends discusses the moral rights of animals or the responsibilities that firms, governments and individuals should assume in the context of the current climate emergency. Additionally, I identify three types of moral agency (expressive, reflective and ethical) according to the functional capacities exerted by agents at each level. After introducing this framework, I will argue that firms meet the necessary conditions to be considered moral agents of these types.

This chapter makes several contributions to our knowledge of CMA. First, it highlights the constitutive role played by communication in morality. The authors of the communicative turn describe morality as a phenomenon that occurs in communication. However, they overlook the constitutive connection between the normative structure and the communicative interactions. In their view, the normative structure emerges and evolves outside interactions. In contrast, I assume that all elements of morality are constituted in and through communication. Neither agents' moral expressions nor normative structures exist prior to or beyond communication. Second, I describe morality as a phenomenon that occurs at various levels, depending on the communicative functions exerted by agents. This conceptualization of morality can be highly productive in a twofold manner. From a theoretical standpoint, it emphasizes the diverse ways in which firms can operate morally. From a practical viewpoint, it could inform decisions made by managers and regulators encouraging firms to develop the necessary competencies to function at these levels.

This chapter is structured as follows. Section 1 introduces its basic elements. Section 2 outlines the communicative turn that has recently emerged in CMA research. Section 3 elucidates the communicative nature of morality, the levels at which it occurs, the required competencies, and the types of moral agency exerted at those levels. Section 4 emphasizes the plausibility of asserting that firms meet the necessary conditions for operating at these levels. Section 5 outlines the main contributions of this chapter and underscores its implications for research and business practice.

2 Corporations and Morality. The Communicative Turn

A relevant issue in business ethics is whether firms can be the object of rights and expectations generally connected to morality. This issue is often framed as a dualistic debate between proponents and opponents of the idea of corporations as moral agents. According to the authors supporting this idea—often called *collectivists*—firms fulfill the necessary conditions, so they should be considered moral agents (e.g., Hess 2018; List and Pettit 2011). Those who reject this idea—often called *individualists*—argue that firms are fictional entities incapable of independent action, and only humans can be considered moral agents (e.g., Sepinwall 2017; Velasquez 2003). In this debate, authors on both sides structure their arguments by addressing two questions: what are the necessary conditions for moral agency?, and do firms meet those conditions? In the most frequent approach, authors address the first question by exploring how morality happens among humans. Based on that description, they address the second question by evaluating whether firms fulfill the conditions previously identified for humans.

In the traditional development of this debate, individualist authors address the first question by highlighting certain phenomenological conditions—such as embodiment or feeling emotions—that humans fulfill, but firms do not (e.g., Mansell 2008; Sepinwall 2017). In contrast, most collectivist authors take as a necessary condition some function that humans exert and we can identify by analogy in firms. Proponents of this approach—whom we could call *functional collectivists*—usually mention as essential conditions the capabilities for intentional action (e.g., Donaldson 1982; French 1979), autonomous decision-making (e.g., Pettit 2007; Hess 2018), and value judgment (e.g., Donaldson 1982; Pettit 2007). These authors interpret moral agency as a trait that is not contingent on a biological or phenomenological condition, but on effectively performing the internal functions required to act with intentionality and moral autonomy.

In recent years, this debate has reached a point where the positions between individualists and collectivists are too far apart for a fruitful dialogue. This situation has prompted some authors to propose alternative arguments supporting CMA. This impulse has given rise to two alternative lines of reflection. On one hand, some authors adopt a pragmatic approach (e.g., Hsieh 2017; Mulgan 2019). For these authors—whom we can call *pragmatic collectivists*—, the practice of attributing moral agency to firms is justified by their utility to society, regardless of whether or not firms meet certain conditions analogous to those of individuals. On the other hand, a small group of authors has recently begun to highlight the roles that firms often play in morality, interpreted as a phenomenon that occurs through communication. In the image of these authors, firms are moral agents because they effectively participate in those communicative interactions in which morality occurs. The most relevant proposals of this emerging communicative turn are diachronic responsibility (French 2017), the conversable firm (Pettit 2017) and, more recently, the second-person account (Morrison et al. 2022). In contrast to other approaches to CMA, the authors of this turn emphasize the necessity of communication for morality. Table 1 summarizes the main currents in the academic debate on CMA. In what follows of this section, I review the proposals mentioned in the communicative turn and assess how they have helped us advance our understanding of CMA.

Diachronic responsibility

French (2017) introduces the notion of diachronic responsibility to revise and update his earlier ideas on CMA. Specifically, he starts from the frequent observation among humans of cases in which we evaluate a person for a past action in a different way if we appreciate that they have undergone an inner change (e.g., by expressing regret or compensating the injured party). Based on this point, French identifies two types of responsibility: synchronic and diachronic. Synchronic responsibility is assessable at the time (T) when the agent performs the action. In contrast, diachronic responsibility is assessable at a later time (T n + 1). Within this framework, an agent will be considered synchronically responsible at time T for an action carried out at that moment if its internal decision structure is causally linked to the action. Furthermore, the agent will bear diachronic responsibility at a later time (T n + 1) if it maintains the internal decision structure that led it to perform the action at time T. French links diachronic responsibility to the agent's structural continuity and states, as a general rule, that diachronic responsibility for a past action will depend on the degree to which

Table 1 Main approaches to corporate moral agency

Theoretical position in relation to CMA	Moral agent	Conditions for moral agency	Prominent examples
Individualists	Individuals	Phenomenological qualities, such as embodiment or feeling emotions	Mansell (2008) Sepinwall (2017) Velasquez (2003)
Decision-based functional collectivists	Individuals corporations	Rationality- and control-based qualities, such as acting intentionally, deciding with autonomy, and making value judgments	Donaldson (1982) French (1979) Pettit (2007) Hess (2018)
Pragmatic collectivists	Individuals corporations	Utility for society	Hsieh (2017) Mulgan (2019)
Communication-based functional collectivists	Individuals corporations	Effective participation in moral communications	French (2017) Pettit (2017) Morrison et al. (2022)

the agent's current decision structure is consistent with the structure that caused the behavior in the past (p. 57).

After introducing this idea, French applies it by analogy to corporations. Consequently, a firm's synchronic responsibility for a past action will depend on the extent to which it has maintained or modified the *internal decision structure* (CIDS) it applied when performing the action. As French notes, "corporate moral responsibility may be enhanced or diminished over time depending on whether there is operational connectedness of its current organizational mechanism (particularly policies and procedures) to the one that was functional at the time of the event" (p. 60). Within this framework, if a firm rectifies the defects of its CIDS that led to the misconduct, its diachronic moral responsibility may decrease over time, or even disappear.

However, French points out that a corporation's CIDS is rarely accessible from the outside. As a consequence, it is usually difficult to assess its diachronic moral responsibility. To resolve this difficulty, French turns his attention back to humans, who express their internal structures through 'self-told narratives' (p. 61). Individuals elaborate autobiographical narratives to reconstruct their identity and present a coherent image of themselves to others. French points to an analogous ability in corporations when they "articulate their narratives in annual reports, in advertising, in legal documents, in internal and external statements of corporate culture, and in policies" (p. 62). According to French, this self-narrative exercise helps any firm to interpret itself and expose itself to external control that prevents it from telling "delusional stories" or including in its narratives "gross errors of significant and publicly observable facts" (p. 63). Within French's framework, communication occurs primarily in a unilateral manner—going from the corporation to the external world—and is only replicated by other agents to prevent delusional deviations in corporate self-narratives. In the final paragraph of his essay, French stresses, however, that the discursive element (self-narratives) is only a non-necessary means to access a company's CIDS. In his words, "narrative coherence issues can be jettisoned if we are satisfied, as I am, that operational connectedness is adequate as a ground for diachronic ownership of past corporate and human actions" (p. 64). Thus, French emphasizes the need to prioritize attention—when attributing moral responsibility to any agent—to its internal decision structure, considering its self-narrative as an indicator that allows us to access that structure.

The conversable corporation

Pettit (2017) offers his view of CMA as a stylized syllogism. Building upon two premises (firms are conversable agents and conversable agents are morally responsible for their actions), he concludes that firms should be held morally responsible. He develops his first premise (firms are conversable agents) in three steps. First, he interprets agency as the capacity of an entity to act. In his view, an entity is considered an agent when it "maintains certain purposes, forms reliable representations of its environment, and acts reliably so as to satisfy its purposes according to those representations" (p. 17). Second, he argues that corporations are agents in this functional sense. Like any other agent—for example, a person, an animal, or a robot—firms may act erratically in anomalous situations, but apart from those situations "we can identify various purposes by which they hold, [...] we can take them to form reliable judgments or representations about the opportunities at hand or the best means of realizing those purposes; and we find that they more or less reliably act so as to satisfy their purposes according to their representations" (p. 18). As in previous work, Pettit identifies the internal decision structure as the mechanism that enables corporations to exercise these capabilities (see, for example, Pettit 2007). From this point, Pettit closes the first syllogism by asserting that firms are conversable agents, insofar as they "are able to speak for themselves and to offer an interpretation of their own purposes, representations, and actions" (p. 19).

When it comes to explaining how conversable agency functions in firms, Pettit asserts that it operates within the firm and also outwardly. In its internal dimension, this agency implies the firm's ability to interpret itself as an agent and reflect, reinforcing its capability to form internal representations, set purposes, and choose alternatives. In its external dimension, this agency is expressed when the firm presents itself to others and interacts. In both dimensions, conversable corporate agency always occurs in a delegated manner, through humans who participate in its governance structures or behave as its spokespersons (pp. 19, 23).

To justify his second premise (conversable agents are responsible for their actions), Pettit relies on the commissive dimension of communication that we as humans often experience. When we express to others our desires and intentions, we commit ourselves to fulfill them and assume the risk of being socially penalized for failure (p. 21). Therefore, being a conversable agent means interpreting and expressing reality and oneself through language. From this point, Pettit closes his syllogism by affirming that firms are fully conversable agents and should therefore be held accountable for their actions (p. 32).

The second-personal account

Morrison et al. (2022) offer what they call a 'second-personal account' of CMA, which they develop in two steps. In the first step, they follow the philosopher Stephen Darwall to explain the moral agency of humans. In the second step, they project this framework onto firms. Specifically, Darwall (2006, 2018, 2021) is concerned with those frequent cases in which some humans express reactive attitudes toward others.[1] In Darwall's view, when an individual expresses a reactive attitude toward another—such as indignation, resentment, or guilt—they face what Darwall calls a 'second-personal address' (Darwall 2021, p. 564) that implies the attribution of a moral responsibility for the recipient. From this idea, Darwall focuses on the capability that the appealed agent has (or does not have) to respond appropriately to the attitudes expressed by others. Darwall refers to that capability as 'second-personal competence' (p. 568), and claims this is a necessary condition for considering any entity as 'an accountable moral agent' (p. 566).

On this basis, Morrison and colleagues point out that the essential activities of morality (i.e., expressing reactive attitudes toward other agents and responding to the reactive attitudes of others) are part of a process that is essentially communicative. Therefore, an agent can be held morally responsible "if and only if that agent has the competence to participate in a particular type of conversation" (p. 326). Such competence requires that the agent meet three conditions: the attitude condition, the direction condition, and the authority condition. An agent fulfills the attitude condition when it makes sense "for other agents to form reactive attitudes toward that agent" (p. 326). Conversely, an agent fulfills that condition if it is able to understand and react to the demands of others. An agent fulfills the direction condition when it is able to participate with others in a dialogue based on reactive attitudes (p. 328). Finally, an agent fulfills the authority condition when it is able to recognize the moral authority of other agents or, to put it another way, when it is able to recognize other agents as valid sources of moral reasons, deliberate about itself independently of its own beliefs and desires, and respond appropriately to the second-person discourses of others (pp. 329–331).

In a second step, Morrison and colleagues apply this conception of moral responsibility to firms. In this case, the authors start from the frequent observation of situations in which corporations are subject to reactive attitudes from individuals. Applying the framework used for individuals, the authors argue that a firm may be considered a moral agent if it meets these three conditions (attitude, direction, and authority). Thus, a firm will be considered morally responsible if it can perceive and internally process the reactive attitudes of other agents (attitude condition), can engage in dialogue with those agents, connecting that dialogue with its internal deliberative processes (direction condition), and can recognize, in that dialogue and deliberations, the other agents as valid sources of moral claims (authority condition). The latter condition is only fulfilled if the firm can consider the intentions of those agents in its internal deliberation, adopt a reflective point of view and correct their behaviors when necessary. Morrison and colleagues introduce a feature of functional relativity in their framework, by claiming that these conditions cannot be recognized indiscriminately in all corporations. Thus, we can only attribute moral responsibility to a firm if it behaves as a suitable object for our reactive attitudes, which in turn depends on its culture and governance system (p. 342).

The communicative turn in CMA research. How far we have come

The reviewed proposals have clear differences. I will point out two of them. First, each proposal highlights a condition for moral agency. French emphasizes the importance of the agent's decision-making structure, which is revealed through its self-narrative. Pettit highlights the agent's ability to deliberate internally and engage with others through language. Morrison and colleagues emphasize the agent's ability to engage in internal deliberation and dialogue with others based on reactive attitudes. Although all these essays emphasize the communicative dimension of moral agency, each interprets the conditions differently. Second, the cited essays diverge about the degree of ontological causality they attribute to communication. In fact, these essays can be categorized based on this criterion. At the most fundamental level—which we may term *accessorily instrumental*—, French interprets the agent's self-referential narrative as a sign ideally dispensable for appreciating the continuity of its decision structure. At an intermediate level—that we may term *necessarily instrumental*—, Pettit interprets discourse as an essential element for morality, but the key condition for moral agency remains internal to the agent, namely, its ability to deliberate and engage with others through language. At a deeper ontological level—that we may call *expressively constitutive*—Morrison and colleagues claim that an entity can be considered a moral agent only when it appropriately reacts to others' moral demands, occurring all this process through communication.

However, these proposals share certain elements that help assess the progress achieved by the communicative turn. First, they all highlight an idea that went unnoticed by other collectivist approaches to CMA (see Table 1). Specifically, they emphasize that morality is an inherently social phenomenon that occurs through communication. This means that we cannot appreciate the moral agency of any entity if we leave communication aside. Second, they interpret morality from a functional perspective. In line with decision-based functional collectivists, the authors of the communicative turn connect moral agency to the practice of certain capabilities. However, the latter link such capabilities to communication. Third, the reviewed essays explore morality among adult and mature humans, and then apply their ideas to firms. This methodology has some ideological implications. By first exploring morality among humans, the reviewed essays assume that human beings are the paradigmatic moral agents and that only humans and firms are legitimate candidates to participate in the moral community.

Taking these ideas into consideration, the reviewed essays represent a remarkable advance over the available approaches to CMA. In contrast to those approaches, these essays emphasize that moral agency always emerges through discourse. However, these proposals have at least two key limitations. First, by attributing moral communication to humans and firms, they avoid considering whether other entities may be considered moral agents. This point may not appear significant when we reflect on CMA. However, if we take human beings as the prototypical agents for morality and focus our efforts on projecting their qualities onto firms, we will ignore how morality actually occurs. As a sign of this deficiency, we won't be able to connect our ideas with some relevant insights about morality that have been recently established in other fields of research, such as biology, philosophy of technology, or computational science. In these fields, an increasing number of authors attribute some type of moral agency to non-human entities, like animals, technologies, and objects (for some relevant examples, see de Waal 2009; Latour 1992, 2002; Martin 2019; Monsó 2017; Verbeek 2008). If we focus on humans and firms, we will fall short of understanding a phenomenon that appears to be ontologically transversal. Second, the reviewed essays do not account for all modalities of moral communication. In this regard, the most advanced proposal is that of Morrison and colleagues. According to them, the process of moral responsibility occurs when an agent is the object of reactive attitudes expressed by others. This view is acceptable, but factually incomplete. In our ordinary lives as moral beings, we easily observe that we often communicate about morality without reference to the behaviors or reactive expressions of others. This observation helps us understand that morality includes certain dynamics that have not yet been connected to CMA. In the next section, I explore those dynamics. To do so, I develop a fully functional approach to morality, whatever the ontological nature of the entities involved.

3 Communication and Morality. A Systems Approach

In this section, I address my first two research questions: how does morality occur through communication and, from my answer to that question, what functions does an entity exercise when it communicates morally? In contrast to the available proposals in the communicative turn to CMA, my point of reference is not morality as it usually occurs among humans, but morality as a phenomenon that emerges and evolves when agents exert certain functions causing certain effects that activate certain processes. Based on this premise, I will examine the functions that any agent must perform for morality to occur. To do this task, I will draw on Luhmann's ideas. In particular, I will explain, correct and extend some of his main ideas on communication and morality. This implies assuming the essential logic of his ideas, but also exploring new paths that will lead us to claims different from Luhmann's. I will highlight where these divergences arise.

3.1 What Is Morality

Morality is a complex phenomenon and can be defined in very diverse ways. Broadly speaking, morality can be interpreted with a normative or a descriptive approach. The first is characteristic of ethics and associates morality with what is good or right. The second is common in social sciences like sociology, economics, or ethnography and describes morality as a phenomenon that occurs in social interactions. In the context of this divergence, Niklas Luhmann assumes the second approach. According to Luhmann, morality is not rooted in the soul or mind of human beings, nor is it something that links humans to transcendent values like good, peace or justice (1992a, p. 996). Morality is one dimension among others of communication and fulfills its function in the sphere of social relations (1996, p. 28). More specifically, Luhmann interprets morality as an artifact that helps individuals coordinate their mutual expectations when they are involved in an interaction and none of the functional codes available in society—such as prices for economics, truth for science or legality for law—is directly applicable to that interaction. In other words, according to Luhmann, morality serves individuals engaged in a stable interaction (or *social system*, in Luhmann's terminology) to evaluate the extent to which their mutual expectations are compatible (see Luhmann 1991, 1992a, 1996).

In Luhmann's social theory (1995, 2002, 2006), when two agents (A and B) are connected in a social system S(A-B) and neither can predict how the other agent will behave, a situation of uncertainty arises for both agents.[2] Luhmann designated this situation as *double contingency* (e.g., 1995, p. 108). In this context, the agents cannot express and coordinate their expectations in a straightforward way. The task is too complex. Therefore, they can only do that in a simplified way. Specifically, they can only coordinate their perspectives using three resources that conform morality: (1) a normative structure that includes elements such as rules, values, principles or examples, (2) a binary code to attribute good/bad values to perceived attitudes and behaviors, and (3) a symbolic index—esteem, with its opposite value, contempt—that each agent uses to express the degree to which they evaluate the other agent's perspective as compatible with their own. In the practice of morality, human agents use these resources simultaneously and in a complementary way. Thus, the binary code of morality enables A to assign a simplified value (good/bad) to B's behaviors based on A's perception of conformity between B's behaviors and the normative structure operating in their system S(AB). Once a binary value (good/bad) is assigned to a behavior, the index of esteem allows A to express whether, from its point of view, the integration of its perspective with that of B is acceptable or unacceptable (Luhmann 1996, p. 30). In a dynamic dimension, the moral resources (normative structure, binary code, and expressions of esteem) enable the agents involved in an interaction to align their perspectives—thus ensuring the continuity of the system—, or to leave the interaction aside, when they appreciate that their perspectives are totally incompatible.

Let us illustrate Luhmann's view of morality with an example. If two young people meet for the first time at a concert and begin chatting about any issue (their musical preferences, the band's best songs, their experiences at other concerts…), they will quickly bond in an emergent system (Luhmann 2006, p. 49; Weick 1977, p. 273). At the end of the concert, they exchange phone numbers and agree to stay in touch. The next day they meet for a chat in a café. As they talk, they realize that a good friendship is developing. At one point in the conversation, one of them proposes to meet on Sunday morning to go to a record fair. When they meet on Sunday, one of them reproaches the other for being late. The angry tone helps the other understand their divergence about punctuality in appointments. In a case like this, moral communication helps the friends to understand to what extent their perspectives are compatible and what behaviors are (or not) acceptable within the boundaries of their system.

Luhmann interprets morality in a functional way, as it serves the agents involved to coordinate their perspectives and manage the continuity of the system. If the appealed friend believes that the reproach for being late is unfair or disproportionate, she will understand that the adjustment of their perspectives is not adequate. If she values the friendship positively, she will attempt to adjust her behavior patterns or persuade her friend to be more tolerant on that issue. However, as Luhmann emphasizes, morality is not always the best mechanism for ordering interactions in functionally differentiated societies. For instance, moral resources are not the most suitable for determining a defendant's guilt in a trial or assessing the validity of a scientific theory. These operations are too complex, and are typically regulated by mechanisms specific to the functional systems. In such situations, morality would generate serious dysfunctions (1992a, p. 1004). For instance, it would be unacceptable for a judge to apply the law with varying severity based on his moral opinion of the defendant or the acts committed.

In Luhmann's theory, morality takes the whole person as a reference (Luhmann 1978). Thus, when a person evaluates another person by applying the resources of morality, they do not take as a reference the specific capabilities of the evaluated person, but their aptitude to participate in the system. When applying moral resources, agents A and B evaluate the compatibility of their perspectives and express their judgments through esteem. Expressing my gratitude to another symbolically indicates the compatibility—from my point of view—of our perspectives about the system. Conversely, blaming or expressing resentment toward another symbolically expresses the mismatch I perceive between our perspectives. For this reason—Luhmann argues—, we should not seek the function of morality in promoting good behaviors instead of others that are intrinsically inadequate. When interpreted through a strictly functional lens, morality is seen as a mechanism that, operating through communication, helps agents manage the uncertainty that arises in their social interactions (Luhmann 1992a, p. 1004; 1995, p. 236).

I accept this conception of morality. However, I disagree with Luhmann on two relevant points. The first has to do with the agents involved in the system, the second with the social space where morality operates. Regarding the agents, Luhmann argues that only humans can engage in moral communication (e.g., 1995, pp. 82, 130, 235). In my view, what counts for attributing moral agency to an entity is its empirically verified ability to engage in moral communication with other agents. This communication can be executed through verbal and symbolic language—as humans often do—but also through other semiotic modes. In the realm of known interactions, agents can communicate morally with nonverbal signs, such as facial gestures, body movements or physical contact. In this sense, primatologist Frans de Waal has explained the proto-moral dynamics that link chimpanzees and other ape species in their social groups (1996, 2009, 2013). Although not having a linguistic code as complex as that of humans, animals of certain species are able to evaluate the behaviors of their peers and communicate their judgments through a wide range of expressions, including reactive attitudes such as gratitude, punishment, regret, forgiveness, or anger (de Waal and Andrews 2022). In the realm of unknown interactions, we cannot rule out the possibility that other semiotic modes (yet to be discovered or invented) serve certain groups of agents to communicate about what is acceptable or unacceptable in their interactions. When we consider morality with a functional approach, we can leave behind that anthropocentric substratum that takes humans—or more precisely, a humanistic image of human beings—as the inevitable reference for morality. From a strictly functional point of view, it does not matter what kind of expressions or linguistic code agents use, nor the ontological category we assign to those agents. What counts is the efficacy with which those agents operate in moral communications.

My second divergence relates to the social space of morality. According to Luhmann, morality always occurs within the boundaries of the social system that binds agents (1995, p. 1004). As a result, each group has its specific moral dynamics that help agents coordinate their perspectives. I partially disagree with this idea. In my view, an agent can asses the desirability of maintaining a social system in which it actually participates, as well as evaluate the potential viability of other system in which it is not actually involved. In the last case, the agent hypothetically evaluates its normative fit with other agents. Expressed another way, an agent A may morally evaluate an agent B in the context of an actual system S(AB) in which the two participate, but A may also evaluate its normative fit with an agent C with whom it is far from interacting. The first case occurs, for example, when one of us complains about being rudely interrupted while speaking during a family dinner. The second occurs when any of us criticizes the war atrocities ordered by a foreign ruler whose abuses we learned about from the news. In both cases—real or hypothetical interactions—esteem serves agents as a symbolic index to express their judgment on a matter that is difficult to address directly: to what extent it is or would be desirable, from their point of view, to maintain a social system with the agent being evaluated.

This idea is supported by observing how we often react to certain practices that were frequent in the past but are now culturally distant from us, such as slavery, cannibalism, child labor, animal cruelty or degrading humor. When considering such practices, most of us express indignation and rejection. In contrast, an anthropologist would approach that practice with a more objective—or, to put it another way, less moral—attitude. Specifically, she would explain what functions those practices had in their societies, how they were carried out, through what rituals... When we position ourselves as moral agents, we tend to feel indignation and symbolically express our refusal to participate in a real or hypothetical system with agents who engage in such behaviors. In contrast, the anthropologist adopts a scientific approach—and, with it, a morally neutral attitude—that allows her to refrain from evaluating the desirability of maintaining a system with those agents.

3.2 *Levels (or Modes) of Moral Communication*

Social systems exist when agents are linked by mutual expectations and communicate with each other. If agents cease communicating and no longer have mutual expectations, the system disappears. If two people who were close friends stop talking to each other and neither counts for the other, the system ceases to exist (Luhmann 1992b, 1995). Despite their dynamism, social systems gain stability as certain interaction patterns emerge. Consider a group of friends who gather every Thursday night to cook and chat over dinner. The flow of past interactions has generated certain patterns that guide the friends in coordinating tasks, selecting topics for conversation, and expressing opinions. We can imagine, for example, that the members of that group often talk about their shared experiences, avoid talking about politics and explain recent anecdotes at work. Social systems are composed of operations—or communicative events—and these operations are strongly conditioned by the normative structure shaped by previous operations (Luhmann 1995). Operations—or communicative events—and structure are mutually connected, but this relationship is not a closed loop. Social agents always communicate in a context of uncertainty—as they do not know for sure how those they are addressing think, how they will interpret and react to their messages or what new stimulus will operate from the environment—, and this uncertainty opens communication to new developments. When two agents communicate, neither of them knows with certainty with what communicative event the other will react to their message. In this situation, each new communicative event will be conditioned—but not determined—by the previous structure of expectations, and that structure will evolve as the agents exchange their messages.[3]

This duality of structure and process is also recognizable in moral communication. However, there is a crucial point to explore regarding the object about which agents communicate when they communicate morally. To delve into this, I address the following question: what do agents communicate about, when they communicate morally? To answer this question, Luhmann identifies two objects. According to Luhmann (1978), morality serves humans to communicate about: (1) the attitudes and behaviors of other agents or (2) the elements—rules, principles, values…—that constitute the normative structure of the system. In other words, agents can communicate morally to evaluate the behaviors or attitudes of other agents, but also to make explicit, criticize, reinforce or reject one or more elements of the normative structure operating in their systems.

I categorize the orientations mentioned above as the expressive and reflective levels of morality. At the *expressive level*, agents communicate their evaluations of other agents' behaviors or attitudes using the symbolic index of esteem. When confronted with the behaviors or attitudes of others, they express their indignation, anger, resentment, gratitude, admiration or any other evaluative emotion that takes as a reference their previous expectations. At the *reflective level*, agents communicate about the normative structure that serves as a reference for their evaluations and expressions of esteem. Comparing the objects of communication at these levels, we observe that, at the expressive level, agents relate behaviors and attitudes to a normative structure. Instead, at the reflective level, agents problematize the normative structure that they usually take for granted in their interactions. When one friend expresses anger toward the other for being late, moral communication occurs in an expressive mode. If the reproached friend responds that he was in a traffic jam, that the reason is not serious enough to get angry, or that the other has also done things in the past that annoyed him, both friends will still communicate at the expressive level. The same is true if the appealed friend admits his fault and apologizes. However, if both friends start talking about the idea of tolerating such behavior, as long as the latecomer alerts the other with a call or a text message, they would be communicating in a reflective mode. In this case, agents do not react to certain attitudes or behaviors. Instead, they communicate explicitly about certain structural elements on which their reciprocal judgments of esteem depend.

Both modes of moral communication fulfill different functions. The communicative events that occur in the expressive mode allow the social system to continue functioning, dynamically coordinating the expectations and behaviors of the agents through their interactions. Instead, the events that occur in the reflective mode allow agents to problematize elements of the normative structure supporting the system. The elements that served as a reference for expressive morality now function as an object of interest and discussion. This dynamic enables the system to reflect on itself and, as a consequence, to correct its internal inconsistencies and evaluate its alignment with the broader environment. Let us imagine that one day, one of the group's friends explains during dinner that, in his opinion, some jokes told in the group are offensive to a collective to which he feels strongly connected. After discussing the matter, the friends agree to refrain from making such jokes. By communicating in this way, the friends make the normative structure of the group the explicit object of their collective reflection. Through that communication process—and regardless of whether or not they reach agreement—the system structure is reintroduced into the dynamics of the system as an object for communication. The group of friends has thus been able to perform a valuable exercise of self-reflection, evaluating the adjustment between its internal conditions and the environment.

Transcending Luhmann's ideas, I identify a third level for moral communication. Sometimes agents adopt an abstract an general approach to communicating about one or more elements of the normative structures that should regulate interactions. For instance, several individuals may talk about the normative values and principles that everyone should prioritize in the case of animal entertainment shows or crimes committed by adolescents. Agents communicating in this way adopt a metamoral perspective that I refer to as *ethical level*. The reflective and ethical modes are similar in the dialogical and non-evaluative attitude assumed by agents: they communicate about the structural elements of morality, rather than about others' behaviors and attitudes. However, those modes maintain an important difference. Thus, agents communicate in a reflective mode when they express their opinions about the normative structure that guides, or could guide, their actual or plausible interactions, and they communicate in an ethical mode when they assume a non-participatory attitude to express their opinions about how interactions in any system should be coordinated.

We typically associate this type of communication with debates among academics in the field of Ethics. Following the logic of systems theory, when communication occurs in that field, each communicative event (such as the publication of an article in a journal, or the presentation of a paper at a congress) is expected to be linked to previous events in the communicative tradition of the field, and to make a sufficiently relevant contribution to stimulate further communicative events. However, if we interpret ethics as a mode of communication that simply involves a non-participatory attitude, we observe that this type of communication often occurs between agents not professionally engaged in academic dialogues. In other words, agents often communicate ethically without acting as ethics scholars.

In this section, I have explained how social agents engage in morality by communicating about different things. In presenting my view, I have drawn on Luhmann's ideas, from which I have diverged at various points. My aim has been to transcend Luhmann's assumption that morality and communication are limited to humans. I have also added an ethical level to those described by Luhmann. To achieve this, I have offered a strictly functional approach to communication and morality. Table 2 illustrates the main similarities and divergences between Luhmann's view and my own. In the following section, I will describe the capabilities that an agent must exert to operate at each level of morality.

Table 2 Comparing functional approaches to morality

	Luhmann's approach	A communicative and strictly functional approach
Similarities		
Nature of morality	Communicative, normative and socially restricted to systems	
Function of morality	Assess compatibility and coordinate perspectives among agents	
Resources for morality	Normative structure, binary code (good/bad) and esteem as an index	
Differences		
Phenomenon to be explained	How morality occurs among humans	How morality occurs
Social scope for morality	Existing social systems	Existing social systems and those hypothetically conceivable
Moral agents	Human beings	Entities with functional capabilities for moral communication
Levels (or modes) of morality	Expressive and reflective	Expressive, reflective and ethical

3.3 *Modes of Moral Agency*

In the previous section, I explained that morality operates at three different levels, depending on the object of communication. I summarize them briefly. At the expressive level, agents offer or receive expressions that symbolically indicate the appreciated alignment of their behaviors with the normative structure of the system. At the reflective level, they dialogue with a participatory attitude about the normative structure of the system. At the ethical level, they assume a non-participatory attitude to communicate about any structural element of morality. In a functional view, morality is a heterogeneous phenomenon, and agents need to exercise different abilities to communicate at each level. From this point, I identify three types of moral agency according to the level at which agents communicate.

First, I attribute *expressive moral agency* to any entity that is capable of expressing esteem to others and reacting to others' expressions of esteem, taking those expressions as an index to symbolically assess the compatibility of their reciprocal expectations. This type of agency involves two capabilities or conditions. The first is demonstrated in specific interactions, when an entity expresses or reacts to others' expressions of esteem. The second is recognizable over time, when an entity's behaviors show that it has internally considered others' expressions of esteem to correct its behavioral patterns. In a functional view, the conditions for any type of agency strictly depend on the agent demonstrating the effective exercise of certain capabilities through practice. Applying this rule, I refer to the former capability as an *episodic expressive condition*, and the latter as a *dynamic expressive condition*.

The distinction between both capabilities (or conditions) is relevant. Thus, we can only say that an entity behaves as an expressive moral agent when it demonstrates its ability to express and react to the esteem expressions of others, while also indicating through its behaviors its internal capability to readjust its decision-making structure. In the realm of known interactions, we can appreciate this full expressive moral agency in most humans, including young children, but also in members of some animal species. As I mentioned earlier, many biologists and philosophers have already attributed this type of agency to species like killer whales, dogs, elephants or primates (e.g., Monsó and Wrage 2021). It is now widely recognized that certain animal species exhibit reactive attitudes of approval or rejection to the behaviors of their community members (episodic expressive condition) and adjust their behaviors in reaction to the esteem expressions of those with which they interact (dynamic expressive condition). In the realm of unknown but plausible interactions, we cannot rule out the possibility that certain non-human entities (yet to be discovered or invented) effectively manage esteem expressions to adjust their understanding of what is acceptable or unacceptable in their interactions.

Second, I attribute reflective moral agency to any entity that is capable of communicating with other entities about any element of the normative structure operating in the social system in which they participate. I refer to this capability as a *reflective condition*. In using this capability, the entity problematizes some element of the normative structure it shares with others in a system. The first two types of agency (expressive and reflective) have a restrictive relationship. This means that not all expressive moral agents can operate at the reflective level. We can thus affirm that expressive moral agency operates at the lowest level of morality—without the term 'low' implying a negative value—and thus functions as a basis (or necessary, but not sufficient condition) for reflective agency.

Third, I attribute ethical agency to any entity capable of communicating with other entities about any attribute of morality, regardless of how that attribute manifests in the normative structure of the systems in which that entity participates. I refer to this capacity as an *ethical condition*. Compared to the other types, ethical agency is exerted in communications that do not refer to the systems that involves the communicating agents. As far as we know today, this type of agency can only be practiced by mature adult humans. However, its functional identification allows us to conceive—at least as a plausible hypothesis—the possibility of non-human entities engaging in theoretical dialogues about morality. Imagine, for example, that an artificial agent is able to engage in reasoned communication with a human—or another artificial agent—about the practical rules to be applied in quelling a street protest, minimizing harm to civilians in war, or resolving the ethical dilemmas associated with self-driving cars. In any of these cases, the artificial entity would be acting as an ethical agent, receiving and/or providing non-participatory arguments to make a theoretical reflection on morality, without necessarily being programmed to express or react to esteem.

After presenting my approach to moral agency, several points need clarification. First, the types of moral agency are not directly attributed to any ontological category. Although these types are easily attributed to humans, we cannot rule out the

Table 3 Modes of moral communication and moral agency

Level of morality	Object of communication	Type of moral agency	Conditions for moral agency
Expressive mode	Behaviors and attitudes	Expressive agency	Offering and reacting to esteem expressions/correcting behaviors in reaction to those expressions
Reflective mode	Normative structure of the system	Reflective agency	Capability to communicate about the normative structure of the system
Ethical mode	Attributes of morality	Ethical agency	Capability to communicate about morality with a non-participatory attitude

possibility of them being exerted by other types of entities, whether existing or yet to be invented. Second, the attribution of any type of agency is always done through observation. In my strictly functional view, what really matters for attributing agency to an entity is not its internal mechanisms, but the capabilities it exerts in interactions. In this sense, moral agency operates a similar way to other socially conferred attributes, like reputation, legitimacy, or trust. Being qualities of an entity, they are attributed by others. Third, the described levels for moral communication may come into contact, generating interferences. In other words, the communicative dynamics occurring at one level may cause certain resonances in the dynamics developing at other levels. For example, the ideas that one agent expresses at the ethical level may be interpreted as offensive by another agent, that reacts with indignation. The point I want to emphasize is clear: the objects of communication that are generated at one level of morality can be reintroduced as an input for communication at other levels. This possibility of reintroductions and interferences does not invalidate the explanatory value of my framework, which becomes more valuable as it helps us understand how morality occurs. Table 3 summarizes my conception of morality.

4 Corporations as Discursive Moral Agents

In the previous section, I described the three levels (or modes) of moral communication and the functions an entity must perform to operate at each level. Building on this view, I now address my last research question: is it plausible to consider corporations moral agents? To answer this question, I will explore whether it is appropriate to say that firms fulfill the necessary functions to operate effectively at each level of morality. As an intermediate step, I will first explain how firms act and communicate.

How firms act and communicate

Firms are social entities created by humans to operate in business. It is an undisputed fact that firms cannot speak, think, touch or experience emotions as humans do. Corporations cannot express anger, admiration or contempt toward other agents.

Nor can they participate directly in public debates to express their opinion on any issue. Despite all these limitations, firms act and communicate effectively, using other agents as intermediaries. Firms act and communicate in a mediated way. In the field of organizational studies, scholars from the Montreal School—as a research line within the *communication constitutes organization* approach, or simply CCO—,[4] have extensively studied and explained this phenomenon (e.g., Cooren 2006; Taylor and Robichaud 2004). In particular, these scholars argue that organizations engage in interactions in a mediated way, through other agents (such as a human spokesperson, a building, a sign on a wall, a financial report, an advertisement…) that act or communicate on their behalf. Within this stream, Cooren (2010) introduced the concept of *tele-action* to describe how an organization can make a difference in a situation by acting and expressing itself through other agents. According to Cooren, mediation is a key aspect of communication, as all entities communicate through an intermediary (be it a text, a sound, a photography…) that makes them present (p. 28). In the case of corporations, it is clear that they can only communicate through other entities that express on their behalf (p. 74).

In communicative processes, the concept of tele-action (or mediated action) is complemented by the notion of *social attribution of agency*. If action in general—and expression in particular—, always occurs in a mediated way, agency (or authorship) is inherently shared. Adopting a flat, relational ontology inspired by Latour (2005, 2013), Cooren emphasizes that the authorship of anything we encounter in our world (a text, a decision, a scientific theory, a language…) is the result of a network of mediations that distribute agency among a plurality of entities (Cooren 2018). From this perspective, he argues that no entity (human or non-human) communicates or acts independently, and authorship is always a shared attribute. Regarding communication, what we express in any situation is always the result of a series of mediations in which we use other entities as mediums. For example, when conversing with others, our expression is inevitably mediated by our vocal organs, the air carrying the sound and the physical distance between us and our listener. Complementarily, other entities (ideas, languages, feelings…) utilize us as a means to manifest themselves in a given situation. Therefore, we can assert that communication—as any other form of action—always unfolds within an extensive network of mediations and influences involving a potentially unlimited number of agents. The inherently distributed nature of action—particularly expression—means that the observer has to choose to which agent in that immense network they attribute authorship of the considered action (Cooren 2006, p. 87, 2015, p. 4).

In the case of firms, the phenomena of mediation and attribution of agency have been described and empirically validated by CCO scholars. In many of their works, these authors have shown that utterances made by certain agents on behalf of a corporation (such as human speakers, signs, documents…) produce relevant effects in the firm's social context, being those utterances and effects socially validated and attributed to the firm (Christensen et al. 2013). The key point of this approach—which I subscribe to in its essential premises—is that the attribution of an expression to a firm does not depend on whether we can make a clear causal connection between any internal mechanism of the firm (such as having a deliberative structure or a strong corporate culture) and the expressed message, but rather on whether that message has been accepted and attributed to the firm by other agents. What really matters in assessing the agency of a collective entity is not the functional effectiveness of certain internal attributes, but the recognition and validation by others of what any other agent has said or done on behalf of that entity. In summary, we can say that a firm always acts and communicates in the middle of an extensive network of mediations, and within that network, the firm will only appear as the author of an action or message when the agents it interacts with recognize its agency (Dobusch and Schoeneborn 2015; Schoeneborn et al. 2019). In the following sections, I will build upon this insight to assess whether it is plausible to claim that firms—at least some firms—can behave—at least in some situations—as moral agents.

4.1 Corporations as Expressive Moral Agents

I defined above an expressive moral agent as any entity that can express esteem to other entities and react to others' expressions of esteem, taking those expressions as an index to symbolically evaluate the compatibility of their perspectives. This type of agency involves the capability to manage esteem expressions in specific interactions (episodic esteem condition) and to alter the agent's pattern of behavior in reaction to those expressions (dynamic esteem condition). We usually attribute both capabilities to corporations and other collective agents. For instance, we recognize the first capability when we observe that a firm reacts on social media by apologizing for any behavior that was interpreted by others as inappropriate or offensive. We recognize the second capability, for instance, when the firm reacts to complaints received from its customers or employees by modifying some of its commercial or hiring policies. While the first capability manifests through a specific communicative event (such as publishing an apologetic message), the second one manifests through a longer sequence of events attributed to the firm. In such situations, we usually accept that corporations can express and react to esteem expressions from others, and also manage those expressions to assess the degree of alignment, or compatibility, between their behavioral patterns and the expectations of others.

Graza's recent crisis illustrates how expressive moral agency operates for corporations. Graza is a successful U.S. startup that sells squeezable bottles of extra virgin olive oil. In December 2022, many customers complained to the firm because their orders had arrived late, poorly packaged, or damaged (Cohen 2023). Some failures could be attributed to the firm, while others were related to heavy snowstorms that disrupted transportation in many regions of the country. In response to those complaints, Graza's CEO, Andrew Benin, sent an e-mail to the 35,544 customers who had ordered products during the previous 60 days. In his email, Benin expressed an apology, acknowledged that many customers had good reasons to be disappointed, explained the causes of the failures, and offered all email recipients—who far outnumbered those who experienced the problem—a $4.43 discount coupon for future orders. From a formal perspective, the message was 835 words long, included several misspellings, and alternated the singular pronoun 'I' and the plural pronoun 'we'. Specifically, Benin used the singular pronoun 10 times, in sentences like "I want to apologize for not being excellent in terms of your customer experience" or "I can understand how receiving your Sizzle with an apparent 'dent' or 'compression' can be frustrating," and the plural pronoun 30 times, in sentences like "we are aware that we may have disappointed [error included] some folks over the past 60 days" or, in an phrase repeated four times, "we are sorry". The message functioned as a sincere expression of apology that could be attributed to Benin as well as to the firm. In the following days, Graza received more than 800 responses from customers, expressing their gratitude to the firm. Some wrote "Thanks for your honesty, I wish more businesses did the same", or "I won't be using the discount, but I will be reordering" (Bell 2023). Whatever the internal mechanism applied at Graza to react to complaints (such as a crisis committee formed by several managers, advice from external experts, a CEO personal decision…), customers attributed the message to Graza and, with it, the expressive moral agency. The key point was that this attribution was expressed in subsequent communications by customers who interacted with the firm.

In Graza's case, we can appreciate how expressive moral agency operates for companies. Prior to the crisis, the firm engaged in stable interactions with many clients, in which both parties had formed mutual expectations about their behaviors. Although certain individuals or teams made decisions at the firm, customers attributed authorship of those decisions to the firm—that is, they viewed Graza as an agent—and they conceived their expectations toward the firm. In December 2022, when Graza's products did not arrive as expected, many customers perceived a mismatch between their expectations and the behaviors attributed to the firm. That mismatch led many customers to express their disappointment, taking the firm as their point of reference. Within a few days, Graza resumed acting and communicating as an agent, taking the received complaints as an index of the disappointment felt by its customers. Through a message signed by its CEO—but whose material authorship was unknown—the firm behaved as an expressive moral agent. It did not matter who actually wrote that message or for what reasons. It did not matter whether that decision was personally made by the company's CEO, agreed upon by a crisis committee, drafted by an artificial agent (like ChatGPT) or mechanically executed by a middle manager applying a previously designed crisis management plan with a template for apology messages. In any of these plausible scenarios, what really matters is that the majority of customers viewed the firm as an expressive moral agent, demonstrating this understanding in their subsequent communications.

In the organizational literature, several lines of research support the expressive moral agency of corporations. In the strategic field, several authors describe how firms often bind themselves to other agents in stable interactions, using esteem expressions to coordinate their expectations and behaviors (e.g., Dowling 2016; Ham and Kim 2019). As an example, Bosse et al. (2009) claim that, among firms engaged in strategic relationships, "boundedly self-interested behaviors' are much less frequent than economists assume, as the agents involved in an interaction are usually 'bounded by norms of fairness" (p. 448). As these authors explain, firms tend to behave in an intensely communicative way, rewarding agents for actions they consider fair and incurring costs to punish agents engaging in behaviors they consider unfair. Additionally, a robust stream of research on corporate crises describes the positive effect of firms' apologies, when interpreted as sincere, in restoring trust among interacting agents (e.g., Racine et al. 2020; Yakut and Bayraktaroglu 2021). Abundant literature on CSR has also demonstrated that expressions of criticism and rejection directed at firms often operates as a mechanism to redirect their behavior (e.g., Matejek and Gössling 2014; Schultz and Seele 2020). As an example of this trend, Noack et al. (2019) demonstrate how appropriate actions by a firm following criticism can facilitate the recovery of its eroded legitimacy.

4.2 Corporations as Reflective Moral Agents

I stated above that an entity behaves as a reflective moral agent when it communicates about elements included in the normative structure of an interaction in which it participates. I referred to this capability as reflective condition. In contrast to the expressive mode, when an entity exerts this capability, it problematize some element of the normative structure of the system. The normative elements that are taken for granted in the expressive mode now become an object of communication. We often attribute this capability to firms, for example, when we observe several members of its management team negotiating with workers' representatives to change some of the firm's working policies, or when we read about a corporation signing an agreement with an NGO to improve its environmental policies. In these and many other similar cases, we acknowledge that firms—acting through their representatives—exert the necessary aptitude to communicate with others about the normative structure operating in their relationship.

The interaction between Amazon and its employee group Glamazon illustrates well how this type of agency operates for corporations. In recent years, Amazon has repeatedly stated its commitment to certain principles and values connected to genre and sexuality (Amazon 2024). To design policies that fulfill its commitment, the firm has recognized various employee groups as key stakeholders. In particular, the group Glamazon represents the interests of LGBTQIA+ employees and community. Building on this recognition, the management team negotiates with Glamazon representatives to design the firm's policies that affect that collective. During these interactions, representatives speak on behalf of both groups about the normative structure within their system. Both entities recognize each other as a reflective moral agent. Illustrating this dynamic, a group of Glamazon members and Amazon's HR department collaborated in 2016 to design a 'toolkit' to assist transgender employees in their transition process (Bird 2018). That toolkit included a FAQ section, a template for communicating transition in the workplace, and a best practices guide on issues such as privacy, official records, pronoun usage, and dress within the firm. During this process, both entities interacted reflectively through their representatives to negotiate and agree on certain normative elements that should guide their behaviors.

However, reflective dynamics do not always lead to agreement. In March 2021, several hundred Amazon employees signed an internal petition to remove the book *Irreversible Harm* from the firm's platform arguing that it violated the firm's policies by framing LGBTQIA+ identity as a mental illness. The firm's senior management rejected the demand, justifying its decision by its policy of tolerating diverse viewpoints, "even when those viewpoints differ from our personal viewpoints or Amazon's stated positions" (Kim 2022). In the following weeks, Glamazon representatives held several meetings with members of the firm's management team. During these meetings, Glamazon representatives made several demands. They asked to change the category of the book to 'conservatism' instead of 'transgender rights and expression', to label the book as 'misinformation', to donate the book's profits to a trans-focused charity and to make the decision process more transparent. Amazon's management team reacted to these demands by expressing the firm's support to the LGBTQ+ community, but refused to take these actions, citing the firm's commitment to free speech (O'Donovan 2022). In this case, both entities interacted as reflective moral agents, failing to reach an agreement on the normative structure governing their behaviors.

In organizational research, several streams have supported the idea of firms as reflective moral agents. Thus, many authors have explored the dynamics of dialogue and agreement that connect a corporation with its stakeholders (customers, suppliers, employees…) to establish norms and principles on issues that these agents consider normatively relevant (e.g., Schrage and Gilbert 2021; Sulkowski et al. 2018). As an illustrative example, Burchell and Cook (2013) explained how the dynamics of dialogue between firms and NGOs can lead to relevant transformations in corporate commitments. Adopting a different approach, Olabisi et al. (2019) describe how an indigenous community in Africa transitioned from its position as a non-stakeholder for a firm—thereby lacking leverage power and legitimacy—to the status of a major stakeholder. Analyzing an under-attended phenomenon, these authors describe how a collective entity can increasingly gain reflective moral agency, strengthening its capacity to dialogue about the normative structure that guides agents in their interaction.

4.3 Corporations as Ethical Agents

As I said above, an entity behaves as an ethical agent when it communicates with a theoretical and non-participatory attitude about how any structural element of morality should be. I referred above to this capability as ethical condition. We usually attribute this capability to corporations and other collective entities. We do so, for instance, when we observe a senior executive of a firm participating in a forum where a set of standards are being negotiated for all agents in an industry, or when we see a firm publishing a message online criticizing the behavior of another agent (a government, a political group, another firm…), appealing to a value or principle whose normative force extends beyond the boundaries of any particular social system. In situations like these, we tend to accept that a corporation communicates about what is or should be considered appropriate, without reference to the normative structure of any specific system in which that firm participates.

The recent case of caste-based discrimination in high-tech corporations illustrates well how this type of agency operates. The issue became popular in 2020, when the California labor regulator sued Cisco Systems over the case of a lower-caste engineer who accused two upper-caste bosses of blocking his career (Moreno and Smith 2022). Since then, caste-based discrimination in technology firms has been the object of intense communications among activists, politicians, journalists and many other groups, all emphasizing the ethical imperative of eradicating this form of discrimination in business. Throughout this communication process, some high-tech firms declined to comment on the issue or denied that their internal dynamics involved such discrimination. Conversely, other corporations stepped forward. Apple updated its employee code of conduct, explicitly prohibiting discrimination based on caste (Dave 2022). Amazon also expressed this prohibition through its institutional commitment to human rights (Amazon 2023).

The reactions of Apple and Amazon had a dual effect on moral communications related to the issue. When a firm openly expresses its support for a general normative principle—in this case, equal opportunity concerning caste—the firm's statement operates at the ethical level of moral communication. The corporation communicates its support for a normative element to which it attributes a validity that extends beyond the boundaries of all its interactions. According to the firm, 'caste equality' is a principle that all agents should respect in any interaction. However, this expression also has a potential effect on the firm's specific interactions with other agents. Regarding commitment, an agent's ethical discourse has a strong performative dimension (Christensen et al. 2013). Thus, when a firm declares its support for a specific ethical element—in this case, equal opportunity concerning caste—it is highly probable that many of the agents with whom the firm interacts (employees, activists, customers…) will take those statements into account for their expectations, introducing them at the expressive level of their communication with the firm. For example, if an activist group appreciates an inconsistency between the firm's behavior and its general statement, it is likely that this group will express its moral indignation toward the corporation. Moral life is complex, levels of moral communication are connected, and a firm's ethical statements often affect the expressive mode of its moral communications (for an example of this interference affecting McDonald's, see Regalado 2020).

In the organizational literature, several lines of research support the characterization of firms as ethical agents. In this vein, many authors have emphasized that firms often participate in forums where diverse actors (firms, governments, local associations, NGOs…) discuss complex issues and generate—through dialogue, negotiation and engagement—normative structures to manage them. For instance, Scherer, Palazzo and colleagues assume a Habermasian approach to explain that transnational corporations play a crucial role in global governance, driving or participating in the creation of shared normative frameworks that promote and guarantee the rights of many agents (e.g., Scherer and Palazzo 2007; Scherer et al. 2006, 2013). As those authors explain, these normative and shared frameworks establish what agents consider appropriate or inappropriate in a much higher and more general way than the specific limits of their interactions. Supporting this idea, some authors describe actual cases of corporations engaging with other agents (authorities, competitors, NGOs, local communities…) in deliberative processes to negotiate the rules and procedures that should be applied by any agent regarding an issue (e.g., Høvring et al. 2018; Scherer and Voegtlin 2020). These authors support the idea that firms can discuss and negotiate certain normative elements whose scope is intended to extend far beyond the boundaries of their specific interactions.

5 Final Discussion

In this chapter, I offer a functional conceptualization of CMA, interpreted as a phenomenon that always and strictly happens through communication. In developing this conceptualization, I characterized morality as a phenomenon that always occurs in the context of social interactions and may develop in different ways. According to this perspective, morality is neither a simple nor a homogeneous phenomenon, and the same applies to moral agency. To develop this idea, I described morality as a communicative phenomenon that occur at three levels, depending on the object that agents communicate about. At the expressive level, they evaluate behaviors through esteem. At the reflective level, they communicate about the normative structure that coordinates their interactions. At the ethical level, they communicate with a non-participatory attitude about the abstract attributes of morality. Each of these levels is connected to a specific capability and thus to a certain type of moral agency, which I identified as expressive, reflective and ethical. Whether an entity exerts one type or another does not directly depend on its internal attributes, but on its capability—recognized by the agents with whom it interacts—to adequately exert the functions required for each level of moral communication. Within this general framework of moral agency—which is applicable to agents of any ontology—, CMA is only a specific case.

My proposal assumes and extends some crucial insights of the communicative turn to CMA (see Sect. 2). In line with this turn, I connect CMA to communication. However, my view differs from available proposals on two key points. First, according to proponents of that turn, morality operates among humans and, only by analogy, among corporations. In contrast, I initially reflect on the much more abstract and generic issue of what morality is, how it operates among agents of any ontology, and what role it plays in social interactions. Instead of analyzing morality among humans, I first reflect on morality as a cross-cutting phenomenon, without presupposing what kinds of entities participate in it. Building on this foundation, I explore the functions that any entity must exert to behave morally. Subsequently, I argue that it is plausible to recognize such functions in corporations.

Secondly, I argue that moral agency strictly depends on an entity's capability to effectively communicate with others about certain moral objects. On this basis, I accept the possibility that entities of diverse ontology apply different mechanisms—some already known, others yet to be discovered or invented—to engage in morality. In my framework, what matters is an agent's effectiveness in communicating morally, regardless of the mechanism employed. In my view, what matters is the observed behavior of a firm, its ability to stimulate new communications form other agents validating the recursivity of its message. What matters is its ability to prompt more communication, rather than the internal mechanism that allows an agent to communicate in some way. This suggests that we should not arbitrarily disregard or prioritize any mechanism of firms to exercise their moral functions.

This approach extends the communicative turn, in at least two ways. The first relates to the modes for moral communication and the second to the type of entities that can be recognized as moral agents. Regarding the modes for moral communication, the available proposals within the communicative turn describe certain dynamics that occur at the expressive level. Despite their differences, these proposals emphasize the moral relevance of the reactive expressions of others toward the firm. I identify two additional levels for morality. Regarding the entities that can be recognized as moral agents, I move beyond the traditional dualistic approach that has restricted moral agency to humans and firms. In my framework, morality is a phenomenon open to the participation of entities with diverse ontologies, some with ontologies already known and others with ontologies yet to be discovered or invented.

Finally, I propose two lines for future research. First, this essay identifies three levels of moral agency. Although this distinction is analytically valuable, we often observe that moral dynamics simultaneously occur at several levels. I have referred to these interferences in Sect. 3.3. It would be valuable to investigate in depth how these interferences occur, how they are amplified or mitigated. Second, this essay should encourage the development of a cross-cutting theory of morality. In organizational studies, morality is often interpreted as a phenomenon that occurs exclusively among mature, adult humans. This anthropocentric and exclusionary view has been eroded in recent history by incorporating non-human entities—such as animals, corporations, or intelligent artificial systems—into the moral community. Each of these additions is relevant, but the development of a theory of morality that characterizes it functionally and as an ontologically cross-cutting phenomenon remains underdeveloped. In my opinion, such a theory should focus on the discursive functions that agents fulfill when operating morally. The conceptualization of morality proposed in this essay may serve a suitable starting point for constructing that theory.

Notes

1. For the theoretical root of this approach, see Strawson (1962).
2. It is important to emphasize what social systems are made of. According to Luhmann (1992b, 1995), social systems are made of communications or, to put it more precisely, they are made of communicative events that are recursively and self-referentially linked. This means that each communicative event links to a previous event and is at the same time the stimulus for the next one. If the communicative events cease, the system disappears. This idea implies that any

phenomenon or entity other than these recursively linked communicative events is considered part of the system's environment, including the communicating human beings.

3. Morality does not necessarily mean reciprocity. The elements of a normative structure do not have to prescribe the same behaviors for all the agents involved in a system. In a family system, for instance, the normative structure may not establish identical expectations for all members. What matters is that the expectations—whether symmetrical or not—are known and shared by all members of the system.
4. For a recent review of the CCO approach, see Bencherkiet al. (2022).

References

Amazon (2023). *Amazon Global Human Rights Principles.* Amazon website. Available at: https://sustainability.aboutamazon.com/society/human-rights/principles.

Amazon (2024). *Amazon leadership principles.* In Who we are (Amazon website). Available at: https://www.aboutamazon.com/about-us/leadership-principles.

Bell, C. (2023). Mastering the art of strategic apology. *Forbes.* Available at: https://www.forbes.com/sites/chipbell/2023/02/14/mastering-the-art-of-strategic-apology/?sh=b90471676 51a.

Bencherki, N., Basque, J., & Kuhn, T. (2022). Introduction. In J. Basque, J., N. Bencherki & T. Kuhn (Eds.). *The Routledge handbook of the communicative constitution of organization.* New York: Taylor & Francis.

Bird, J. (2018). *Amazon UK launches transgender toolkit to support employees. Employee Benefits.* Available at: https://employeebenefits.co.uk/issues/june-2018/amazon-uk-transgender-toolkit/.

Block, K. (2024). Major win: 100% of McDonald's eggs in the US now come from cage-free hens. *The Humane Society of the United States.* Available at: https://www.humanesociety.org/blog/Mcdonalds-cage-free-eggs-us.

Bosse, D. A., Phillips, R. A. & Harrison, J. S. (2009). Stakeholders, reciprocity, and firm performance. *Strategic Management Journal, 30*(4), 447–456.

Burchell, J. & Cook, J. (2013). Sleeping with the enemy? Strategic transformations in business–NGO relationships through stakeholder dialogue. *Journal of Business Ethics, 113*, 505–518.

Cain, A. (2022). Carl Icahn slams McDonald's for putting out 'hollow' responses to 'animal welfare violations'. *Business Insider.* Available at: https://www.businessinsider.com/carl-icahn-slams-mcdonalds-animal-welfare-violations-2022-4.

Christensen, L. T., Morsing, M. & Thyssen, O. (2013). CSR as aspirational talk. *Organization, 20*(3), 372–393.

Cohen, B. (2023). What happened when the olive-oil startup apologized. *The Wall Street Journal.* Available at: https://www.wsj.com/articles/sorry-graza-olive-oil-apology-11673476845.

Cooren, F. (2006). The organizational world as a plenum of agencies. In F. Cooren, J. R. Taylor and E. J. V. Every (Eds.) *Communication as organizing: Empirical and theoretical explorations in the dynamic of text and conversation.* Mahwah, NJ: Lawrence Erlbaum, 81–100.

Cooren, F. (2010). *Action and agency in dialogue.* Amsterdam, ND: John Benjamins.

Cooren, F. (2015). In medias res: Communication, existence, and materiality. *Communication Research and Practice, 1*(4), 307–321.

Cooren, F. (2018). Materializing communication: Making the case for a relational ontology. *Journal of Communication, 68*(2), 278–288.

Darwall, S. (2006). *The second-person standpoint: Morality, respect, and accountability.* Cambridge, MA: Harvard University Press.

Darwall, S. (2018). "Second-personal morality" and morality. *Philosophical Psychology, 31*(5), 804–816.

Darwall, S. (2021). Recognition, second-personal authority, and nonideal theory. *European Journal of Philosophy, 29*(3), 562–574.

Dave, P. (2022). Caste in California: Tech giants confront ancient Indian hierarchy. *Reuters*. Available at: https://www.reuters.com/business/sustainable-business/caste-california-tech-giants-confront-ancient-indian-hierarchy-2022-08-15/.

de Waal, F. (1996). *Good natured: The origins of right and wrong in humans and other animals*. Cambridge, MA: Harvard University Press.

de Waal, F. (2009). *The age of empathy: Nature's lessons for a kinder society*. New York: Harmony.

de Waal, F. (2013). *The bonobo and the atheist: In search of Humanism among the primates*. New York: Norton & Company.

de Waal, F. B., & Andrews, K. (2022). The question of animal emotions. *Science, 375*(6587), 1351–1352.

Dobusch, L. & Schoeneborn, D. (2015). Fluidity, identity, and organizationality: The communicative constitution of Anonymous. *Journal of Management Studies, 52*(8), 1005–1035.

Donaldson, T. (1982). *Corporations and morality*, Englewood Cliffs, NJ: Prentice-Hall.

Dowling, G. R. (2016). *Winning the reputation game*. Cambridge, MA: MIT Press.

French, P. A. (1979). The corporation as a moral person. *American Philosophical Quarterly, 16*(3), 207–215.

French, P.A. (2017). The diachronic moral responsibility of firms. In E. W. Orts and N. C. Smith (Eds.), *The moral responsibility of firms*. Oxford, UK: Oxford University Press, 53–65.

Ham, C. D. & Kim, J. (2019). The role of CSR in crises: Integration of situational crisis communication theory and the persuasion knowledge model. *Journal of Business Ethics*, 158: 353–372.

Hess, K. M. (2018). Does the machine need a ghost? Corporate agents as nonconscious Kantian moral agents. *Journal of the American Philosophical Association, 4*(1), 67–86.

Høvring, C. M., Andersen, S. E. & Nielsen, A. E. (2018). Discursive tensions in CSR multi-stakeholder dialogue: A Foucauldian perspective. *Journal of Business Ethics*, 152, 627–645.

Hsieh, N. H. (2017). The responsibilities and role of business in relation to society: Back to basics?. *Business Ethics Quarterly, 27*(2), 293–314.

Kim, E. (2022) Leaked documents show Amazon employees are upset over the company's response to its LGBTQ communities. Now they're organizing a 'stop hate' movement. *Business Insider*. Available at: https://www.businessinsider.com/amazon-employees-are-upset-about-lgbtq-controversies-2022-3.

Ladak, A. (2023). What would qualify an artificial intelligence for moral standing?. *AI and Ethics*, 1–16.

Latour, B. (1992). Where are the missing masses? The sociology of a few mundane artifacts. In W. E. Bijker, and J. Law (Eds.) *Shaping technology/Building society: Studies in sociotechnical change*, Cambridge, MA: MIT Press, 225–58.

Latour, B. (2002). Morality and technology: The end of the means. *Theory, Culture & Society, 19*(5-6), 247–260.

Latour, B. (2005). *Reassembling the social: An introduction to Actor-Network Theory*. London, UK: Oxford University Press.

Latour, B. (2013). *An inquiry into modes of existence: An anthropology of the moderns*. Cambridge, MA: Harvard University Press.

List, C. & Pettit, P. (2011). *Group agency: The possibility, design, and status of corporate agents*. Oxford, UK: Oxford University Press.

Luccioni, A., & Bengio, Y. (2020). On the morality of artificial intelligence. *IEEE Technology and Society Magazine, 39*(1), 16–25.

Luhmann, N. (1978). Soziologie der Moral. In N. Luhmann & H. Pfürtner (Eds.). *Theoritechnik und Moral*. Frankfurt, Germany: Suhrkamp, 8–116.

Luhmann, N. (1991). Paradigm lost: On the ethical reflection of morality. Speech on the occasion of the award of the Hegel Prize 1988, *Thesis Eleven, 29*(1), 82–94.

Luhmann, N. (1992a). The code of the moral. *Cardozo Law Review, 14*, 995–1009.

Luhmann, N. (1992b). What is communication? *Communication Theory, 2*(3), 251–259.

Luhmann, N. (1995). *Social systems*. Stanford, CA: Stanford University Press.

Luhmann, N. (1996). The sociology of the moral and ethics. *International Sociology, 11*(1), 27–36.
Luhmann, N. (2002). How can the mind participate in communication? In W. Rasch (Ed.), *Theories of distinction: Redescribing the descriptions of modernity*. Stanford, CA: Stanford University Press, 169–186.
Luhmann, N. (2006). System as difference. *Organization, 13*(1), 37–57.
Mansell, S. (2008). Proximity and rationalisation: The limits of a Levinasian Ethics in the context of corporate governance and regulation. *Journal of Business Ethics, 83*(3), 565–577.
Martin, K. (2019). Ethical implications and accountability of algorithms. *Journal of Business Ethics, 160*(4), 835–850.
Matejek, S. & Gössling, T. (2014). Beyond legitimacy: A case study in BP's "green lashing". *Journal of Business Ethics, 120*, 571–584.
Monsó, S. (2017). Morality without mindreading. *Mind & Language, 32*(3), 338–357.
Monsó, S., & Wrage, B. (2021). Tactful animals: How the study of touch can inform the animal morality debate. *Philosophical Psychology, 34*(1), 1–27.
Moreno, J. E. & Smith, P. (2022). Rare caste bias case advances, raising calls for Federal action. *Bloomberg Law*. Available at: https://news.bloomberglaw.com/daily-labor-report/rare-caste-bias-case-advances-raising-calls-for-federal-action.
Morrison, A. D., Mota, R. & Wilhelm, W. J. (2022). Relationships, authority, and reasons: A second-personal account of corporate moral agency. *Business Ethics Quarterly, 32*(2), 322–347.
Mulgan, T. (2019). Corporate agency and possible futures. *Journal of Business Ethics, 154*(4), 901–916.
Noack, D., Miller, D. R. & Smith, D. (2019). Let me make it up to you: Understanding the mitigative ability of corporate social responsibility following product recalls. *Journal of Business Ethics, 157*, 431–446.
O'Donovan, C. (2022). Amazon employees protest the sale of books they say are anti-trans. *The Washington Post*. Available at: https://www.washingtonpost.com/technology/2022/06/01/amazon-trans-pride-month-books-protest/.
Olabisi, J., Kwesiga, E., Juma, N. & Tang, Z. (2019). Stakeholder transformation process: The journey of an indigenous community. *Journal of Business Ethics, 159*, 1–21.
Palao, E. (2021). Social animals and the potential for morality: On the cultural exaptation of behavioral capacities required for normativity. In J. De Smedt & H. De Cruz (Eds.) *Empirically Engaged Evolutionary Ethics*. Cham, Switzerland: Springer, 111–134.
Pehn, P. (2023). Nike urged to support supply chain workers. *Ecotextile News*. Available at: https://www.ecotextile.com/2023072130960/social-compliance-csr-news/nike-urged-to-support-supply-chain-workers.html.
Pettit, P. (2007). Responsibility incorporated. *Ethics, 117*(2), 171–201.
Pettit, P. (2017). The conversable, responsible corporation. In E. W. Orts and N. C. Smith (Eds.), *The moral responsibility of firms*. Oxford, UK: Oxford University Press, 15–33.
Racine, M., Wilson, C. & Wynes, M. (2020). The value of apology: How do corporate apologies moderate the stock market reaction to non-financial corporate crises?. *Journal of Business Ethics*, 163, 485–505.
Regalado, N. (2020). McDonald's employees are working through the COVID-19 pandemic. So why don't they have access to paid sick and family leave? *ACLU*, Available at: https://www.aclu.org/news/racial-justice/mcdonalds-is-hiding-policies-that-perpetuate-systemic-racism-behind-woke-washing.
Reuters (2024). Zara faces pressure to publish its full supply chain like other retailers. *Fast Company*. Available at: https://www.fastcompany.com/91054702/zara-inditex-investors-supply-chain.
Scherer, A. G. & Palazzo, G. (2007). Toward a political conception of corporate responsibility: Business and society seen from a Habermasian perspective. *Academy of Management Review, 32*(4), 1096–1120.
Scherer, A. G. & Voegtlin, C. (2020). Corporate governance for responsible innovation: Approaches to corporate governance and their implications for sustainable development. *Academy of Management Perspectives, 34*(2), 182–208.

Scherer, A. G., Palazzo, G. & Baumann, D. (2006). Global rules and private actors: Toward a new role of the transnational corporation in global governance. *Business Ethics Quarterly, 16*(4), 505–532.
Scherer, A. G., Palazzo, G. & Seidl, D. (2013). Managing legitimacy in complex and heterogeneous environments: Sustainable development in a globalized world. *Journal of Management Studies, 50*(2), 259–284.
Schoeneborn, D., Kuhn, T. R. & Kärreman, D. (2019). The communicative constitution of organization, organizing, and organizationality. *Organization Studies, 40*(4), 475–496.
Schrage, S. & Gilbert, D. U. (2021). Addressing governance gaps in global value chains: introducing a systematic typology. *Journal of Business Ethics, 170*(4), 657–672.
Schultz, M. D. & Seele, P. (2020). Business legitimacy and communication ethics: Discussing greenwashing and credibility beyond Habermasian idealism. In J. D. Rendtorff (Ed.), *Handbook of business legitimacy: Responsibility, ethics and society*. Springer, 655–669.
Sepinwall, A. J. (2017). Blame, emotion, and the corporation. The moral responsibility of firms. In E. W. Orts and N. C. Smith (Eds.), *The moral responsibility of firms*. Oxford, UK: Oxford University Press, 143–166.
Strawson, P. F. (1962 [2008]). Social morality and individual ideal. In P. F. Strawson, *Freedom and resentment and other essays*. London, UK: Routledge, 29–49.
Sulkowski, A. J., Edwards, M. & Freeman, R. E. (2018). Shake your stakeholder: Firms leading engagement to cocreate sustainable value. *Organization & Environment, 31*(3): 223–241.
Taylor, J. R. & Robichaud, D. (2004). Finding the organization in the communication: Discourse as action and sensemaking. *Organization, 11*(3), 395–413.
Velasquez, M. (2003). Debunking corporate moral responsibility. *Business Ethics Quarterly, 13*(4), 531–562.
Verbeek, P. P. (2008). Obstetric ultrasound and the technological mediation of morality: A post phenomenological analysis. *Human Studies, 31*(1), 11–26.
Weick, K. E. (1977). Enactment processes in organizations. In B. M. Staw and G. R. Salancik (Eds.), *New directions in organizational behaviour*. Chicago, IL: St. Clair Press, 267–300.
Wrage, B. (2022). Caring animals and care ethics. *Biology & Philosophy, 37*(3), 18.
Yakut, E. & Bayraktaroglu, A. G. (2021). Consumer reactions to product recalls: the effects of intentionality, reputation, and public apology on purchase intentions. *Journal of Business Economics, 91*(4), 527–564.

Is Applying Ethics During a Transitional Period Ethical? A Review of Ethical Leadership and Its Influence on Employees and the Organization

Paulo Viana and Carolina Feliciana Machado

Abstract The development of organizational culture, followers' well-being, and overall performance have all been found to be significantly influenced by ethical leadership (EL). This chapter provides an insight over EL from two angles: first, by outlining its core ideas and characteristics, and second, by highlighting its importance and significance in organizational transition. This paper aims to elucidate how this leadership style can foster a culture of transparency and integrity inside businesses, reduce burnout, and foster trust by examining important theories, definitions, and research on EL. It also has to do with how moral leaders handle the difficult duties of creating a moral atmosphere while becoming used to changing standards and demands. This analysis reveals that EL maintains its ethical and philosophical integrity even during times of transitional organizational change.

Keywords Ethical leadership · Ethics · Leadership · Organizational transition · Employees · Organization

1 Introduction

This chapter is intended to provide up-to-date insight on ethical leadership (EL) based on several articles published in scientific journals (Journal of Business Ethics, The Leadership Quarterly, Journal of Organizational Behavior, British Journal of Management, Academy of Management Journal) providing an in-depth analysis divided into two sub-chapters. The first sub-chapter on how it is to lead ethically and analyzing its impact on employees strategically into three fields: personal (employee's well-being), social (integrity), and economical (boosting performance,

P. Viana · C. F. Machado (✉)
School of Economics, Management and Political Science, University of Minho, Braga, Portugal
e-mail: carolina@eeg.uminho.pt

C. F. Machado
Interdisciplinary Centre of Social Sciences (CICS.NOVA.UMinho), University of Minho, Braga, Portugal

C. Machado (ed.), *Ethics in Management and Business*, Management and Industrial Engineering, https://doi.org/10.1007/978-3-031-82054-0_5

reducing turnover and absenteeism). The second sub-chapter is intended to integrate ethics during a transitional process of change. The positioning of a leader applying an ethical leadership style during such a process and the follower's concern about ethics under a stressful and controversial period are discussed and supported by existing literature.

In contemporary organizational studies, EL has emerged as a central concept, gaining traction as a style that not only guides organizations toward success but also aligns leaders with societal values of integrity and moral responsibility. As Cullen (2022) posited, leading ethically sits at the heart of "brighter" leadership paradigms, providing an alternative to dark or intimidating leadership styles that continuously undermine autonomy and morale and indulge severe directives toward employees. Brighter leadership encourages positive, ethical influences within organizations, shaping environments that support personal and professional development (Bush et al. 2021; Cullen 2022; Lara and Armas 2019). EL's influence is not limited to its inherent moral standards but extends to its impact on organizational culture and employee well-being. Brown and Treviño (2014) highlight that an ethical leader serves as a role model, whose actions inspire followers to emulate ethical behavior. Additionally, ethical leadership has been shown to mitigate employee burnout and absenteeism, thus reducing counterproductive behaviors while enhancing engagement and commitment (Eluwole et al. 2022; Mostafa et al. 2021).

Through a multi-layered approach merging elements of authentic, transformational, and transactional leadership, ethical leaders connect with followers to build and solidify a culture grounded in ethical values and mutual trust. Researchers such as Kalshoven et al. (2020) and Egorov et al. (2020) argue that the uniqueness of EL, including transparency, care for followers' well-being, and honesty, distinguish it from other leadership styles, generating a model where role clarification, power-sharing and fairness are essential components. Consequently, leaders practicing this leadership style, encourage an open and inclusive atmosphere, ensuring that employees feel empowered and truly valued, boosting job satisfaction and reducing turnover intent (Mostafa et al. 2021). Ethical leadership has become crucial in navigating the challenges of organizational change and crisis management. Zahari et al. (2024) highlighted that EL establishes an adaptable and resilient organizational culture that can tolerate internal and external pressures. This perception is reinforced by Badrinarayanan et al. (2019), who posits that EL's emphasis on transparency and moral decision-making induces trust, particularly during periods of transition, thus easing smoother change and fostering morale and loyalty amongst personnel. However, EL's effectiveness and applicability are not without critique. A tension is emphasized within organizations between ethical considerations and operational efficiency, suggesting that, in pursuit of practicality and convenience, organizations may sideline ethical standards during the process of leader' selection (Zyglidopoulos 2021). This ethical gap is more noticeable in high power-distance cultures, where hierarchies may foster environments less favorable to ethical transparency (Blake et al. 2022). Within these ethical complexities, EL encourages for a balance respecting stakeholders' interests and preserves follower's well-being, even within moral dilemmas inherent to leadership (Palanski et al. 2021). Leaders who embrace ethical principles inspiring trust and commitment, foster a culture of integrity driving positive organizational outcomes (Bush et al. 2021; Cullen 2022). This emphasis on ethics becomes significantly pertinent during periods of organizational change, where honesty and transparency are indispensable in conducting employees through often challenging and demanding transitions (Collins and Restubog 2021). Ethical leadership in times of change is about more than just managing employees; it involves leading by example, creating inclusive communication channels, and maintaining the organization's moral compass amid uncertainty. As Payne et al. (2023) note, organizational change is most successful when it occurs within a culture of openness and mutual respect, traits that ethical leaders are uniquely positioned to cultivate. However, ethical leaders face specific challenges, including employee's emotional response to uncertainty, generating resistance, burnout, absenteeism, and detachment. According to Eluwole et al. (2022), by applying ethical principles, these negative outcomes can be mitigated, promoting a path to success, aligned not only to stakeholders, but also to employees well-being and loyalty towards the organization. According to Palanski et al. (2021), ethical leaders are often confronted with ethical dilemmas, balancing their responsibilities to followers with their professional obligations to other stakeholders. This balancing act can be emotionally taxing, as ethical leaders work to uphold values that may occasionally be at odds with immediate organizational demands. However, studies indicate that employees are more likely to reciprocate ethical behavior, trust, and reliability in their leaders, strengthening the organization's moral foundation and fostering a resilient and adaptable workforce (Ahmed and Khan 2023; Badrinarayanan et al. 2019).

2 Ethical Leadership

Whereas there's a dark side, it undoubtedly exists a brighter side. Ethical leadership (EL) is placed on the second pattern due to its unavoidable benefits to the follower and the organization. If, on the one hand, literature on dark leadership focuses on why subordinates feel they must follow a leader, on the other hand, research on brighter leadership focuses on the positivistic impact a leader provides not only professionally but also personally. Hence, the "exemplary models" are defined by Cullen (2022, p. 487).

Among all definitions provided by the literature and in accordance with several authors (Bush et al. 2021; Cullen 2022; Egorov et al. 2020; Lara and Armas 2019; Martin et al. 2022), stating that leaders play a vital part in the process of influencing subordinates regarding ethical and valuable outcomes concerning the organization, the most cited definition of EL, is the one provided by Brown et al. (2005, p. 120) as "the demonstration of normatively appropriate conduct through personal actions and interpersonal relationships, and the promotion of such conduct to followers through two-way communication, reinforcement, and decision-making". However, improvements to Brown et al.'s definition have been made throughout the last decade.

Kaptein (2019) adds moral entrepreneur to the two main components of EL (moral person and moral manager). That is, besides just following current morality, ethical leaders also conduct morality by creating new norms. The author posits that Brown et al.'s (2005) definition needs to be reformulated by adding "the demonstration of normatively appropriate *and new* conduct through personal actions and interpersonal relationships, and the promotion of such conduct to followers through two-way communication, reinforcement, and decision-making" (Kaptein 2019, p. 1146). In addition, the author adds that "ethical leadership does not only follow ethics but that it also leads it" (p. 1136).

Such as other leadership styles, EL has also been stereotyped characteristically. According to Egorov et al. (2020), there are three elements for EL distinction: (1) honesty, reliability, fairness, and concern for follower's well-being. (2) Ethical leaders often apply a transactional style to reward or punish in an attempt to clarify what is expected from the followers (role clarification); (3) Ethical leaders adopt a "giving voice" mentality so that followers express their opinion and feel valuable for decision-making participation (power sharing). In accordance with the previous authors, Kalshoven et al. (2020) add that there are seven dimensions of an ethical leader: **fairness** (integrity and impartiality); **power sharing** (decentralized decision-making process, the leader listens and acknowledges follower's ideas); **role clarification** (entails transparency in communication); **caring and supportive** (follower's trust in a leader if a genuinely approach is displayed); **integrity** (precise and accurate moral principles); **ethical guidance** (awareness and promotion of ethical behaviors); **concern for sustainability** (leader's actions for the environment, promoting recycling programs).

EL is negatively connected to employee burnout, reducing emotional exhaustion, disengagement (Mostafa et al. 2021) and absenteeism (Eluwole et al. 2022). It is believed that EL is an influential method to captivate followers by enhancing positivism and instructing that success is not only about final outcomes but how it is obtained and the path to achieve the final result. By enhancing employee's sense of optimism about the organization's future and their place within, EL acts as a motivator factor to contribute to the organization's success (Badrinarayanan et al. 2019). Hence, encouraging two-way communication, reinforcing trust, and inciting dialogue between members for a relationship of solidarity with each other engaging in helping behaviors enhances the needed bond between leader and follower. Additionally, by acknowledging their fallibility, particularly when they are not required to, leaders induce a stronger relationship in the leader-follower relationship as honesty is an ethical behavior that is normatively appropriate, enhancing employee's ethical leadership evaluation (Zhang et al. 2024).

Ethical leaders have also the power to develop and strengthen relationships, and model behaviors (followers' self-control per se helps ethical leadership to have a more powerful impact on their ethical behavior) (Al Halbusi et al. 2023), focus on procedures and not only results, serving as an inspiration for two-way communication between followers and leaders, providing commitment within the goals, protecting followers from failures and weak performances. When creating a new team for a task, organizations should rely on EL as a style of leadership that transmits confidence shall adversity arises and does not fall back when environments of uncertainty and doubt occur (Martin et al. 2022). EL is, above all forms of positive leadership, hypothetically different. It is a mixture of transformational (role modeling), transactional (punishing or rewarding subordinates for unethical comportment), and authentic leadership (leaders are focused predominantly on transparency and consciousness) comprising a turnover intent's negativity, high levels of job satisfaction and commitment enhancement (Mostafa et al. 2021).

Like many other leadership styles, EL has similarities with several leadership styles. In compliance with participative leadership and in organizations that value ethics, EL engages in a participative environment, inviting subordinates to engage in discussions where their opinion is considered and regarded, and therefore establishing an ethical environment of trust, confidence, and reliability (Yukl 2013). EL can also be considered an essence of inspirational leadership as ethical leaders 'emphasis on ethical values in the organization facilitates subordinates to comprehend leaders 'expectations, providing a clear focus on their duties' ethical aspects and adopting organizational values (Jeong et al. 2024; Zheng et al. 2022). EL is also related to servant leadership not only as a moral consequentialist theory but also because ethical leaders subdue their personal desires and provide willingness, goodness, and astuteness to others (Tan 2024). Authentic leadership is associated with virtue ethics, and deontology is intrinsically associated with EL (Jeong et al. 2024) as the expected behavior from followers complies with established societal standards and norms which are engrained in cultural environments (Amory et al. 2024). According to Badrinarayanan et al. (2019), EL embodies the use of a number of elements from both transformational and transactional leadership styles with a moral dimension, influencing confident behavior and enhancing performance among followers. Besides being considered performers and even "Disneyland-inspired good leaders, a moral peak performer" (Alvesson and Einola 2019, p. 363), ethical leaders dispel these criticisms maintaining trust and confidence in El's practices. In fact, EL is tantamount to honorable, well-educated, and trained leadership (Gollagari et al. 2022).

The outcomes generate concordance among scholars that ethical leaders divide themselves into what concerns morality. That is a moral person (promoting ethical behavior that is assimilated by followers through the observation process) and the moral manager (implementing decision-making, contributing to the mutual dialogue between follower and leader), allowing a citizenship increase and a reduction of counterproductive behavior (Bush et al. 2021). EL has also been linked to a protective style that safeguards followers' interests even when the final outcomes are not the expected or planned ones (Martin et al. 2022). This leads to additional research on EL questioning the true effectiveness of the style. A proactive personality variable was studied and fundament that can substitute EL due to variables such as the follower's ability, expertise, and knowledge along with already established and unified work groups that may not need the traits or characteristics of a leader performing ethically. Moreover, the research found that the lower the self-esteem, the more guidance by ethical leaders is needed by followers; on the other hand, the higher the self-esteem, the less guidance is necessary (Velez and Neves 2018) and proactive subordinates reveal ambition and creativity to pinpoint new ideas for improvement, keep track on personal up to date and are fully aware of company's policies and politics.

Kalshoven et al. (2020) posit that the more ethical leadership is applied, the more ethical the environment becomes, discouraging employee's unethical behaviors such as knowledge hiding and their negative repercussions (i.e., lying and deception) to the leader and the organization (Ali et al. 2024). In fact, EL fosters constructive employee behaviors such as knowledge sharing as it induces a culture of sharing know-how and other valuable information (Sun et al. 2024b) and by sharing information, enhances their prosocial motivation (Xia and Yang 2020). By sharing knowledge contributing to the well-being of all parties, a stronger bond of complicity helps to establish a commitment to achieve personal and organizational objectives. However, should employees perceive that a specific organizational rule or policy is unfair or unjust, they may engage in unethical behavior as a means of rectifying the perceived injustice. This perception of organizational injustice may contribute to the unexpected positive relationship (Ahmed and Khan 2023).

EL provides job satisfaction, commitment, self-efficacy, increases performance, psychological well-being and reduces stress, triggers moral emotions, and consequently avoids the violation of moral conducts and norms (Ahmed and Khan 2023; Jeong et al. 2024). This also contributes to an emulation of ethical leader standards and conduct by the followers as they feel compelled to reciprocate this leadership style (Badrinarayanan et al. 2019; Sun et al. 2024a) due to ethical leaders are seen as trustworthy and reliable role models (Ahmed and Khan 2023; Ali et al. 2024; Eluwole et al. 2022; Jeong et al. 2024; Zheng et al. 2022). Interestingly, Brown and Treviño (2014) have posited that followers who became leaders inspired by ethical role models are more likely to become ethical leaders. EL is also considered an organizationally efficient performance booster as well as an organizational culture enhancer (Alkhadra et al. 2023; Zahari et al. 2024) supported by a leader's ethical attitude towards the organization. According to the same authors, as the project advances, a change in leadership style must be implemented. That is, democratic and participatory styles are recognized as ethical paradigms as leaders share and inspire followers; on the other hand, autocratic styles scarce ethical conduct.

Media plays an important role concerning ethics in organizations. They can be bipolar as information about the organization can propagate unethical behaviors (i.e., payroll injustices), or they can elevate an organization's status quo for several reasons (highly profitable companies, good practices). Brooks et al. (2023) posit that it is not a question of necessity but a matter of survival to ensure organizations act ethically. Due to scandals regarding fraud and corruption in organizations worldwide, pressure from organization leaders to behave ethically has increased significantly (Babalola et al. 2019). Therefore, as organizations fear media, a noticeable increase in discouraging unethical behavior has been raised. Organizations tend to reduce unethical behaviors by implementing a growing number of legal measures (Amory et al. 2024) as an augment of focus on ethics helps to avoid misconduct and can similarly institute a constructive strategy to enhance organizational competitive advantage (Treviño and Nelson 2019). In addition, the more integrated and participative the relationship between the leader-follower is, the more trustworthy EL is. That is, besides promoting well-being and contributing to the performance of employees (Amory et al. 2024) it generates a stronger bond between parties. On the other hand, in cultures where power distance is high, corruption and disloyalty increase, whereas where power distance is low, it is more comfortable for followers to control decision-making from leadership and to get involved when misbehaviors occur, complying with a two-way communication as a behavioral component that is widely related within the definition of EL (Blake et al. 2022; Brown et al. 2005; Bush et al. 2021; Egorov et al. 2020; Velez and Neves 2018). According to Zyglidopoulos (2021), organizations face a huge obstacle regarding leadership as good leadership is considered to be effective and ethical. However, the author acknowledges that even though effectiveness and ethics are needed, organizations tend to select leaders for their efficiency, downplaying ethical needs. Controversially, Collins et al. (2023), state that whenever an internal-intentional crisis caused by the leader arises in the organization (i.e. fraud), the leader is immediately replaced with a leader strongly correlated with ethics.

Interestingly, to what concerns to sharing information, not all employees require the same amount of information as some prefer sociable leaders, whereas others prefer quitter or succinct leaders. If, on the one hand, an ethical leader over-communicates, is in constant contact, and continuously provides updates on employee performance, can come across as pedantic, overloading followers. On the other hand, although some employees prefer, if an under-communication leadership approach is perceived, it can become highly dysfunctional as adopting a *laissez-faire* style can increase psychological distress in followers (Flynn and Lide 2023). This raises a controversial paradigm as applying ethics in an organization requires power-sharing, evolving employees, and requiring their opinions. Assuming that leadership is partially built in the mind of the follower, a leader's ethical intent may not be suitable for the follower's conception of EL (Keck et al. 2020).

3 Applying Ethics and Lead During a Process of Change

Nowadays, organizations face social and psychological issues (among other factors, i.e., economic, and political), and a sense of trust and reliability is deposited on leadership to cope with such issues, even acknowledging that change is not a one-off isolated event, but rather an unceasing and volatile process of positioning and re-positioning an organization to its changing environment. If change is a planned event, it is distinguished by recognizing the need for change in a participative, open, and ethical change process (Payne et al. 2023). Although the private sector has made significant advances in implementing ethical work culture and leadership to foster sustainability and governance, the public sector has not (Zahari et al. 2024). This can contribute to, during a transitional environment from a public to private organization, questions may arise during the implementation of ethics. If followers rely on a leader's guidance on how to interpret organizational change, leaders who provide transparency and honesty (i.e., ethical), about change's impact and its repercussions, are likely to be seen as reliable and trustworthy (Collins and Restubog 2021).

If transactional leadership is to be implemented in a public environment whereas transformational and ethical leadership reigned, the new leadership, by embedding new patterns of management, might not be ethical (Rothausen 2023). An ethical leader may feel confronted when a subordinate does not require his/her leadership, generating a personal and organizational conflict. Shall this conflict persist, organizations will have to adopt "effective strategies to minimize these adverse consequences" (p. 672). In accordance, Banks et al. (2023) suggest that leading in an ethical manner is particularly challenging due to the complexity, ambiguity and uncertainty nature of moral dilemmas in modern organizations. As EL is based on a leader-follower relationship, the more span of control, the more difficult it becomes to practice this style due to the distance between the leader and followers (Jensen et al. 2023). Collins and Restubog (2021), add that, enhanced citizenship and organization member adaptivity are other benefits of EL that need to be addressed as leaders with high levels of impulsiveness and trait anger are improbable to ethical, undermining perceptions of their ability to adapt to organizational change.

According to Payne et al. (2023), the most necessary element for successful change management is EL. Even considering that change by itself is hard, as it induces resistance followed by controversy, uncertainty and insecurity, Payne et al. (2023) posit that if it is managed ethically, the organization can successfully benefit from the new management as it enhances performance and can contribute to a better fulfillment of economical as it can be used as an economic impulse representing a competitive advantage (Crossley et al. 2023), legal and moral responsibilities. According to Badrinarayanan et al. (2019), EL is particularly important during the change process as leaders come under strain and scrutiny; consistently demonstrating EL's practices helps to increase employees' trust, confidence and decrease uncertainty.

Ethical leaders influence the expectations and aspirations of new employees boosting employee's ethical behaviour, commitment and motivation. However, the quest to become an ethically outstanding person and leader entails a lifelong process of moral self-cultivation (Tan 2023) and followers play a vital role in this process as they need to be receptive to ethical ideologies because ethical leaders, to be considered agreeable, they are more likely to endorse appropriate conducts to subordinates (Blake et al. 2022).

A powerful ally of EL is the culture itself. If an ethical culture prevails in an organization, an if leaders have the arduous task to lead the process of change, obstacles such as mistrust and uncertainty will be easier to overcome in an atmosphere of pursuing the ethically correct methodology. EL largely contributes to creating an ethical organizational culture, prioritizing ethical values and behaviors creating an environment where ethics become rooted in the organization's norms and values (Ahmed and Khan 2023). For Zahari et al. (2024), ethical culture is a critical component of organizational culture, encouraging innovative cultures and managing divergent behaviors and not only ethical but also unethical conduct.

However, obstacles arise whenever a new management arises. A considerable timeframe must be considered to learn the new culture and assimilate the cultural characteristics of the organization: this will not ease the indoctrination of the newcomer (Payne et al. 2023). Also, new management should consider hiring employees with enhanced emotional stability who effectively handle challenges and negative emotions and thus are less inclined to experience emotional exhaustion (Sun et al. 2024b) to the existent ones, the new management should consider cultivating a culture of trust that values the reciprocation of help and knowledge sharing (Sun et al. 2024b). O'Keefe et al. (2023) believe that organizations should hire followers who score lower on right-wing authoritarianism as they are more flexible and adaptable to comply with values and norms provided by EL and the organization per se.

Ethical leaders should be considered role models with high ethical standards, setting the rules through a transactional style and providing rewards and punishments to promote employee's moral conduct (Sun et al. 2024a). Through transformational guidelines (by transforming the values of subordinates) (Busch et al. 2023), they can also provide a negative effect such as close monitoring, that is, in order to ensure followers are complying with moral principles and standards of the organization, ethical leaders rely on strict monitoring. However, when such is perceived by followers, they become anxious and anxiety is generated (due to fear of not complying with moral standards that have been imposed) questioning moral principles and conduct (Sun et al. 2024a). In addition, shall the leader engage in a "top-down manner" behavior, an environment of distrust is raised by followers towards the leader, blocking and discouraging bonds between both parties (Liao and Chun 2016). Also, questions arise about the true effectiveness of EL. That is, if, on one hand, ethical leaders act as motivators and role models toward followers, on the other hand, leaders face ethical dilemmas by owing ethical obligations to stakeholders. They act as a weight balance measuring two plates with similar weights as, on one plate, the professional obligations towards stakeholders and the other plate, the moral obligations towards followers. This undoubtedly generates emotional costs of ethical leader behavior as it involves making hard decisions, benefiting some but simultaneously coming at the cost of others (Palanski et al. 2021).

4 Conclusion

Exploring ethical leadership (EL) illuminates its multifaceted and complex role within organizational contexts, particularly in directing organizations through change while promoting an optimistic and constructive ethical workplace. EL goes beyond authentic, transactional, and transformational leadership, interlacing ethical models with psychological, moral, and cultural factors that impact the broader organizational fabric and followers (Badrinarayanan et al. 2019; Brown et al. 2005; Cullen 2022). Research reveals that ethical leaders not only demonstrate appropriate conduct but also act as moral entrepreneurs, recurrently creating new norms that align with evolving ethical standards (Kaptein 2019). By leading ethically, not only influence is transmitted to followers but rather a contagious transformation of the moral landscape with the organizations is perceived, inspiring knowledge sharing, sense of trust, and psychological well-being, while alternatively, reducing absenteeism and burnout (Mostafa et al. 2021; Sun et al. 2024a, b). While culture can shape leadership, leaders also have the power to reshape culture (Hussain et al. 2018), and during the applicability of a culture centered on ethics, EL provides leaders with tools to address the process of change and its challenges, encouraging transparency, honesty and collaboration that ease the transition experience during times of doubt and uncertainty (Payne et al. 2023). Additionally, by supporting participative channels of communication, ethical leaders emphasize bonds of trust, actively reduce resistance, facilitate commitment, and effectively decrease resilience (Collins and Restubog 2021).

On the other hand, there are intrinsic tensions with EL, particularly as leaders strive for the ethical dilemma of balancing stakeholders' interests with followers' interests and obligations within the organization (Palanski et al. 2021). These tensions can be lessened but not completely eliminated; leaders have the onus to cultivate ethical adaptability, considering professional responsibilities against moral imperatives without compromising on ethical values. According to Babalola et al. (2019) and Tan (2024), leaders must not only embody ethical standards but must also have emotional resilience to handle the strain of ethical decision-making, acknowledging that such responsibilities come with emotional costs.

The relationship between ethical leadership and transitional periods of organizational change is multifaceted and profound. During times of change, uncertainty arises, stress increases, and resistance emerges, presenting an opportunity for ethical leaders to guide organizations through these challenges while endorsing moral and ethical principles (Payne et al. 2023). According to Collins and Restubog (2021), this type of leadership becomes inevitable during a transitional process of change, as employees seek guidance, consistency, and reassurance that their organization is not only adaptable but also ethically committed to both its people and objectives. Ethical leaders function as stabilizing agents during transitional periods, reinforcing core values that give employees a sense of continuity and purpose even as the environment shifts (Martin et al. 2022). By consistently modeling ethical behavior, leaders provide employees with a reliable example, demonstrating that despite the changes, the organization remains committed to its foundational values (Badrinarayanan et al. 2019). Additionally, when an ethical leader incorporates fairness and accountability into the decision-making process, the ethical behavior model is instigated, providing an opportunity to followers to emulate it. This consistency reassures employees that, despite the insecurities, uncertainties, and significant costs associated with change, decisions will be made with integrity and transparency, fostering a culture of fairness and collective ethical commitment (Zyglidopoulos 2021). Ethical leaders who acknowledge the difficulties inherent in change and openly discuss challenges, contribute to a culture of empathy and understanding, benefiting employees to cope with transitional periods (Al Halbusi et al. 2023). By involving employees in discussions and decisions regarding the change process, ethical leaders provide to employees a feeling of inclusion where they feel valued and respected, significantly enhancing their commitment to the organization and consequently embracing the transitional period confidently, neglecting fear and uncertainty (Egorov et al. 2020; Velez and Neves 2018).

References

Ahmed, M., & Khan, M. I. (2023). Beyond the universal perception: Unveiling the paradoxical impact of ethical leadership on employees' unethical pro-organizational behavior. *Heliyon, 9*(11), 1–16. https://doi.org/10.1016/j.heliyon.2023.e21618

Al Halbusi, H., Ruiz-Palomino, P., & Williams, K. A. (2023). Ethical leadership, subordinates' moral identity and self-control: Two- and three-way interaction effect on subordinates' ethical behavior. *Journal of Business Research, 165*, 1–18. https://doi.org/10.1016/j.jbusres.2023.114044

Ali, M., Usman, M., Khan, M. A. S., Shafique, I., & Mughal, F. (2024). "Articulating Cognizance About What to Hide What not": Insights into Why and When Ethical Leadership Regulates Employee Knowledge-Hiding Behaviors. *Journal of Business Ethics, 190*(4), 885–895. https://doi.org/10.1007/s10551-023-05426-9

Alkhadra, W. A., Khawaldeh, S., & Aldehayyat, J. (2023). Relationship of ethical leadership, organizational culture, corporate social responsibility and organizational performance: a test of two mediation models. *International Journal of Ethics and Systems, 39*(4), 737–760. https://doi.org/10.1108/IJOES-05-2022-0092

Alvesson, M., & Einola, K. (2019). Warning for excessive positivity: Authentic leadership and other traps in leadership studies. *The Leadership Quarterly, 30*(4), 383–395. https://doi.org/10.1016/j.leaqua.2019.04.001

Amory, J., Wille, B., Wiernik, B. M., & Dupre, S. (2024). Ethical leadership on the rise? A cross-temporal and cross-cultural meta-analysis of its means, variability, and relationships with follower outcomes across 15 years. *Journal of Business Ethics*, 1–29. https://doi.org/10.1007/s10551-024-05632-z

Babalola, M. T., Bligh, M. C., Ogunfowora, B., Guo, L., & Garba, O. A. (2019). The Mind is Willing, but the Situation Constrains: Why and When Leader Conscientiousness Relates to Ethical Leadership. *Journal of Business Ethics, 155*(1), 75–89. https://doi.org/10.0.3.239/s10551-017-3524-4

Badrinarayanan, V., Ramachandran, I., & Madhavaram, S. (2019). Mirroring the Boss: Ethical leadership, emulation intentions, and salesperson performance. *Journal of Business Ethics, 159*(3), 897–912. https://doi.org/10.0.3.239/s10551-018-3842-1

Banks, G. C., Ross, R., Toth, A. A., Tonidandel, S., Mahdavi Goloujeh, A., Dou, W., & Wesslen, R. (2023). The triangulation of ethical leader signals using qualitative, experimental, and data science methods. *The Leadership Quarterly, 34*(3), 1–14. https://doi.org/10.1016/j.leaqua.2022.101658

Blake, A. B., Luu, V. H., Petrenko, O. V, Gardner, W. L., Moergen, K. J. N., & Ezerins, M. E. (2022). Let's agree about nice leaders: A literature review and meta-analysis of agreeableness and its relationship with leadership outcomes. *The Leadership Quarterly, 33*(1), 1–23. https://doi.org/10.1016/j.leaqua.2021.101593

Brooks, S., Richmond, J., & Blenkinsopp, J. (2023). Applying a lens of temporality to better understand voice about unethical behaviour. *Journal of Business Ethics, 188*(4), 681–692. https://doi.org/10.1007/s10551-023-05505-x

Brown, M., & Treviño, L. (2014). Do role models matter? An investigation of role modeling as an antecedent of perceived ethical leadership. *Journal of Business Ethics, 122*(4), 587–598. https://doi.org/10.1007/s10551-013-1769-0

Brown, M. E., Treviño, L. K., & Harrison, D. A. (2005). Ethical leadership: A social learning perspective for construct development and testing. *Organizational Behavior and Human Decision Processes, 97*(2), 117–134. https://doi.org/10.1016/j.obhdp.2005.03.002

Busch, C., Crawshaw, J., Guillaume, Y., & Legood, A. (2023). Ethics-Related Mentoring: A Scale Development and Test of its Role in Promoting Protégé Ethical Behaviour. *British Journal of Management*, 1–21. https://doi.org/10.1111/1467-8551.12707

Bush, J. T., Welsh, D. T., Baer, M. D., & Waldman, D. (2021). Discouraging unethicality versus encouraging ethicality: Unraveling the differential effects of prevention- and promotion-focused ethical leadership. *Personnel Psychology, 74*(1), 29–54

Collins, M. D., & Restubog, S. L. D. (2021). The effect of trait anger and impulsiveness on ethical leadership and support for organizational change. *Journal of Research in Personality, 91*, 1–6. https://doi.org/10.1016/j.jrp.2021.104072

Collins, M. D., Dasborough, M. T., Gregg, H. R., Xu, C., Midel Deen, C., He, Y., & Restubog, S. L. D. (2023). Traversing the storm: An interdisciplinary review of crisis leadership. *The Leadership Quarterly, 34*(1), 1–28. https://doi.org/10.1016/j.leaqua.2022.101661

Crossley, C., Taylor, S. G., Liden, R. C., Wo, D., & Piccolo, R. F. (2023). Right From the Start: The Association Between Ethical Leadership, Trust Primacy, and Customer Loyalty. *Journal of Business Ethics*, 1–19. https://doi.org/10.1007/s10551-023-05485-y

Cullen, J. G. (2022). Moral Recovery and Ethical Leadership. *Journal of Business Ethics, 175*(3), 485–497. https://doi.org/10.1007/s10551-020-04658-3

Egorov, M., Kalshoven, K., Pircher Verdorfer, A., & Peus, C. (2020). It's a Match: Moralization and the Effects of Moral Foundations Congruence on Ethical and Unethical Leadership Perception. *Journal of Business Ethics, 167*(4), 707–723

Eluwole, K. K., Karatepe, O. M., & Avci, T. (2022). Ethical leadership, trust in organization and their impacts on critical hotel employee outcomes. *International Journal of Hospitality Management, 102*, 1–10. https://doi.org/10.1016/j.ijhm.2022.103153

Flynn, F. J., & Lide, C. R. (2023). Communication Miscalibration: The Price Leaders Pay for Not Sharing Enough. *Academy of Management Journal, 66*(4), 1102–1122. https://doi.org/10.0.21.89/amj.2021.0245

Gollagari, R., Beyene, B. B., & Mishra, S. S. (2022). Ethical leadership, good governance and employee commitment: Testing a moderated mediation model in public universities of Ethiopia. *Journal of Public Affairs, 22*, 1–12. https://doi.org/10.1002/pa.2742

Hussain, S. T., Lei, S., Akram, T., Haider, M. J., Hussain, S. H., & Ali, M. (2018). Kurt Lewin's change model: A critical review of the role of leadership and employee involvement in organizational change. *Journal of Innovation and Knowledge, 3*(3), 123–127. https://doi.org/10.1016/j.jik.2016.07.002

Jensen, D. C., Hansen, A.-K. L., Pedersen, L. D., & Andersen, L. B. (2023). Span of Control and Ethical Leadership in Highly Professionalized Public Organizations. *Public Personnel Management, 52*(2), 191–217. https://doi.org/10.1177/00910260221140398

Jeong, S., Sun, C. (Timothy), & Yam, K. C. (2024). The affective processes of ethical leadership: The role of moral emotions. *Journal of Business Ethics*, 1–19. https://doi.org/10.1007/s10551-024-05718-8

Kalshoven, K., Den Hartog, D. N., & De Hoogh, A. H. B. (2020). Ethical leadership at work questionnaire (ELW): Development and validation of a multidimensional measure. *Leadership Quarterly, 22*(1), 51–69. https://doi.org/10.1016/j.leaqua.2010.12.007

Kaptein, M. (2019). The Moral Entrepreneur: A New Component of Ethical Leadership. *Journal of Business Ethics, 156*(4), 1135–1150. https://doi.org/10.0.3.239/s10551-017-3641-0

Keck, N., Giessner, S. R., Van Quaquebeke, N., & Kruijff, E. (2020). When do followers perceive their leaders as ethical? A relational models perspective of normatively appropriate conduct. *Journal of Business Ethics, 164*(3), 477–493. https://doi.org/10.0.3.239/s10551-018-4055-3

Lara, P. Z. M. de, & Armas, M. V. (2019). Does Ethical Leadership Motivate Followers to Participate in Delivering Compassion? *Journal of Business Ethics, 154*(1), 195–210

Liao, E. Y., & Chun, H. (2016). Supervisor monitoring and subordinate innovation. *Journal of Organizational Behavior, 37*(2), 168–192. https://doi.org/10.1002/job.2035

Martin, S. R., Emich, K. J., McClean, E. J., & Woodruff, C. T. (2022). Keeping Teams Together: How Ethical Leadership Moderates the Effects of Performance on Team Efficacy and Social Integration. *Journal of Business Ethics, 176*(1), 127–139

Mostafa, A. M. S., Farley, S., & Zaharie, M. (2021). Examining the Boundaries of Ethical Leadership: The Harmful Effect of Co-worker Social Undermining on Disengagement and Employee Attitudes. *Journal of Business Ethics, 174*(2), 355–368

O'Keefe, D. F., Son Hing, L. S., & Catano, V. (2023). Unethical behaviour in the military: The role of supervisor ethicality, ethical climate, and right-wing authoritarianism. *Military Psychology, 35*(1), 58–75. https://doi.org/10.0.4.56/08995605.2022.2068899

Palanski, M., Newman, A., Leroy, H., Moore, C., Hannah, S., & Den Hartog, D. (2021). Quantitative research on leadership and business ethics: Examining the state of the field and an agenda for future research. *Journal of Business Ethics, 168*(1), 109–119. https://doi.org/10.1007/s10551-019-04267-9

Payne, D., Trumbach, C., & Soharu, R. (2023). The Values Change Management Cycle: Ethical Change Management. *Journal of Business Ethics, 188*(3), 429–440. https://doi.org/10.0.3.239/s10551-022-05306-8

Rothausen, T. J. (2023). Diverse, Ethical, Collaborative Leadership Through Revitalized Cultural Archetype: The Mary Alternative. *Journal of Business Ethics, 187*(3), 627–644. https://doi.org/10.1007/s10551-022-05259-y

Sun, U., Park, H., & Yun, S. (2024a). Ethically treated yet closely monitored: Ethical leadership, leaders' close monitoring, employees' uncertainty, and employees' organizational citizenship behavior. *Journal of Organizational Behavior, 45*(5), 702–719. https://doi.org/10.1002/job.2760

Sun, U. Y., Xu, H., Kluemper, D. H., McLarty, B. D., & Yun, S. (2024b). Ethical leadership and knowledge sharing: A social cognitive approach investigating the role of self-efficacy as a key mechanism. *Journal of Business Research, 174*, 1–15. https://doi.org/10.1016/j.jbusres.2024.114531

Tan, C. (2023). Integrating Moral Personhood and Moral Management: A Confucian Approach to Ethical Leadership. *Journal of Business Ethics*, 1–12. https://doi.org/10.1007/s10551-023-05447-4

Tan, C. (2024). Integrating Moral Personhood and Moral Management: A Confucian Approach to Ethical Leadership. *Journal of Business Ethics, 191*(1), 167–177. https://doi.org/10.1007/s10551-023-05447-4

Treviño, L., & Nelson, K. (2019). Managing business ethics: Straight talk about how to do it right. *Personnel Psychology, 72*(4), 625–626. https://doi.org/10.1111/peps.12360

Velez, M. J., & Neves, P. (2018). Shaping emotional reactions to ethical behaviors: Proactive personality as a substitute for ethical leadership. *The Leadership Quarterly, 29*(6), 663–673. https://doi.org/10.1016/j.leaqua.2018.06.004

Xia, Z., & Yang, F. (2020). Ethical leadership and knowledge sharing: The impacts of prosocial motivation and two facets of conscientiousness. *Frontiers in Psychology, 11*, 1–10. https://doi.org/10.3389/fpsyg.2020.581236

Yukl, G. A. (2013). *Leadership in Organizations Global Edition* (8th Ed.). Pearson

Zahari, A. I., Said, J., Muhamad, N., & Ramly, S. M. (2024). Ethical culture and leadership for sustainability and governance in public sector organisations within the ESG framework. *Journal of Open Innovation: Technology, Market, and Complexity, 10*(1), 1–11. https://doi.org/10.1016/j.joitmc.2024.100219

Zhang, K., Zhao, B., & Yin, K. (2024). When leaders acknowledge their own errors, will employees follow suit? A social learning perspective. *Journal of Business Ethics, 189*(2), 403–421. https://doi.org/10.0.3.239/s10551-023-05329-9

Zheng, Y., Epitropaki, O., Graham, L., & Caveney, N. (2022). Ethical leadership and ethical voice: the mediating mechanisms of value internalization and integrity identity. *Journal of Management, 48*(4), 973–1002. https://doi.org/10.1177/01492063211002611

Zyglidopoulos, S. (2021). On Becoming and Being an Ethical Leader: A Platonic Interpretation. *Journal of Business Ethics, 173*(1), 1–11. https://doi.org/10.0.3.239/s10551-020-04544-y

"Greasing the Wheels of Commerce" or "Sand in the Machine"? The Limits of Traditional Economic Thinking on Bribery in Business and Some Business Ethical Thoughts

Volker Lingnau, Florian Fuchs, and Florian Beham

Abstract Corruption and in specific bribery are as old as human history. Correspondingly, a large amount of academic literature has been focusing on these concepts. Yet, the effects of corruption are still intensely debated. This chapter intends to shed some light on the practice of bribery, exploring the implications from a business ethical perspective. After discussing the main reasons often given for engaging in acts of bribery, we focus specifically on the argument of increasing business performance and the dilemma of market competition, first applying a traditional economic approach. There, we show that the apparently positive effects of bribery are often merely short-term, resulting from underestimating the probability of detection as well as the costs inferred when the bribe is detected. However, and in a second step, such purely economic argument is critically analyzed because it tends to neglect the wider impacts, especially on the legitimacy of businesses. In this context, we emphasize that to sustain such business legitimacy, compliance management is not sufficient. Rather, stakeholder expectations concerning the good conduct of business must be respected. These ethical considerations lead us to conclude that an increased focus on legitimacy management should be considered, transcending simple compliance management approaches.

Keywords Bribery · Gift giving · Compliance · Integrity · Legitimacy · Business ethics

---

V. Lingnau · F. Beham
RPTU School of Business and Economics, University of Kaiserslautern-Landau, Kaiserslautern, Germany

F. Fuchs (✉)
School of Economic Disciplines, University of Siegen, Siegen, Germany
e-mail: florian.fuchs@uni-siegen.de

C. Machado (ed.), *Ethics in Management and Business*, Management and Industrial Engineering, https://doi.org/10.1007/978-3-031-82054-0_6

1 Introduction

Corruption is a widespread phenomenon that is by no means new and not only linked to emerging countries but exists in all areas of the world (Mousavi and Pourkiani 2013). "Corruption has existed and been a problem since the beginning of human history" (Holmes 2015, p. 1). Although corruption and its specific form of bribery is traditionally extensively debated in the context of the public sector (Onkvisit and Shaw 1991; Rose-Ackerman 1997), it is increasingly recognized as a major public concern for private business organizations. In this context, many corruption scandals over the last years illustrate that improper business conduct endangers the trust into entire business sectors (Cleveland et al. 2009). Especially regarding corporate bribery as a specific form of corruption in business, scandals like in the cases of Siemens, Alstom, BAE Systems, Hyundai and Lockheed Martin, have dominated the legal as well as the ethical discussions considering corporate governance issues (Cheung et al. 2012). Although until now, different streams of research have focused on the topic of bribery, there is so far no consensus, neither on definitions nor on the business and societal impact. While some authors state that bribery can be effective and efficient—especially in the short run—for example by acquiring new business contracts (Beck and Maher 1986; Lui 1985) or reducing bureaucratic rigidities (Huntington 1968; Leff 1964), other authors state that bribery behavior can undermine economic growth, strategic competitiveness and in the worst case even threaten the societal "license to operate" of an enterprise (Lingnau and Fuchs 2022). Regarding these two competing perspectives, on the one hand, bribery may appear to pay off and help corporations to generate and increase short-term economic success (Healy and Serafeim 2016) while on the other hand such conduct might threaten the survival of a corporation in the middle or long run if the corrupt behavior is detected (Eberl et al. 2015).

Although there exists a large amount of research on corruption focusing on the legal and individual level, conceptual review and critical business ethical reflections on the organizational impact, especially when transcending purely short-term economistic reflections, are rare so far. Therefore, the following article provides some in-depth reflections on the impact of bribery and in specific an analysis on the justification of corporate bribery, critically scrutinizing traditional arguments in that context. In such vein, we set forth the thesis that the often-emphasized separation between short-term profits and long-term risks of bribery appears plausible on the one hand. However, on the other hand, from a business ethical perspective, such argument seems not sufficient as bribery has much further reaching impacts on the legitimacy of businesses, which needs to be considered from a managerial perspective as well to safeguard the corporation's existence.

2 The Notion of Bribery—Or: When Do Gifts Become a Bribe?

Like the notion of corruption, also the concept of bribery is intensely discussed in research. As Sandholtz and Koetzle (2000, p. 33) rightly emphasize: "Virtually every published work on corruption from the 1960s to the present, wrestles with the problem of defining it". Subsequently, there exists a large amount of literature, however "[d]epending on the individual, ideology, culture, or other context, the term [.] can mean different things to different people" (Luo 2007, p. 213). In the following, we will particularly focus ourselves on the payment of bribes as a specific form of corruption (see e.g., Burke 2016; Gottschalk 2016; OECD 2013). Hence, we will not explore the case of the potentially moral degradation in the broader sense (like waste of public resources, misuse of power, fraudulent actions, etc.), but rather focus on the acceptance and especially on the reflection of granted bribes to business partners as a means to facilitate the acquisition of contracts.

Besides the efforts of establishing a universal definition of bribery, especially the difference to legitimate gift-giving has received some consideration in research. In such manner, Cragg (2018, p. 576) states: "We might say, then, that the story of bribery is a story of the struggle to differentiate gifts from bribes". The question when a gift becomes a bribe is likewise intensely and controversially debated in research and practice. While for example gift-giving on the one side is often acknowledged as certain kind of social interaction in business practice to maintain and build up commercial networks, on the other side, gifts can also create a certain obligation (Dorsch and Kelley 1994; Douglas 2002; Verhezen 2009) at the receiving side. Fritzsche (2005), for example, claims from an ethical perspective that if a gift is intentionally given to influence, it can be regarded as a bribe. Also Argandoña (2005, p. 254) adopts this view and differentiates the two terms by summarizing: "The purpose of a gift is to create an atmosphere of friendship and, as far as possible, make the receiver more inclined to see the giver in a favorable light". While some authors like Fritzsche (2005) and Argandoña (2005) make a clear distinction between a bribe and a gift focusing on the desirability and acceptability of the reciprocal transaction, other authors like for example James (2002) argue that a bribe and a gift cannot be separated clearly. Likewise, also Shaw and Barry (2016, p. 507) state that although business gifts are common and familiar in a global business world "[k]nowing where to draw the line is not always easy". In this context, Fisher et al. (2013, p. 103) even provide a scale of bribery that ranges from "low value gifts on the one side through to holidays, cars and large sums of facilitating payments on the other side" and discuss the acceptability of the different forms.

In this article, we do not intend to solve the discussion and provide a final distinction between a bribe and a gift. Rather, we focus on certain intentions for acts that are definitely regarded as morally or legally impermissible. In general, one can differentiate between active and passive bribery, the first describing the supply side of bribery, while the latter concerns itself with corruptibility on the receiver side (e.g., Gordon and Miyake 2001; Webb et al. 2003). In the following argumentation, we explore the implications for *providing* bribes. Hence, we specifically focus on *active bribery* in demarcation to permissible business gifts from a virtue ethics perspective *as a sufficiently clear illegal or illegitimate intention of the supplier in the act of gift-giving to obtain an unjustified advantage.* In such process, the bribe's supplier puts the own or the corporate interests above those of other parties directly or indirectly affected (stakeholders), who may also have legitimate interests in the business deal (preferred treatment). The concrete appearances of bribery are manifold. As Luo (2005, p. 76) states, bribery can "include 'kickbacks', 'red envelopes', 'gratuities', 'baksheesh', 'grease money', 'facilitation payment', and 'expediting fees'" (see also Tian 2008, pp. 437–438). However, it seems relevant to note that the value of the intended transaction between the two corrupt parties is not always material and in monetary terms (Sanyal 2005). In research, bribing public officials is regarded as the most prevalent form (Shleifer and Vishny 1993). However, bribery is not a single phenomenon in the public sphere and therefore can be found "in any formally organized context in governments and in private firms or in NGOs" (Graycar and Jancsics 2017, p. 1017).

What may be considered as bribery in one culture is not necessarily regarded as an inappropriate behavior in another culture since cultural standards can vary among different countries and even local customs must be considered (Crane and Matten 2016). While some "cultures (e.g., UK, Ireland, USA, Germany, and Switzerland) tend to have more stringent and more elaborate ethical rules pertaining to business gift-giving [, other] cultures (e.g., Japanese, Arabs, Italians, Spanish and Turkish) tend to view the act of gift-giving as an imperative" (Davies et al. 2010, p. 417). Although especially in Western countries, gift-giving in the commercial context is still questioned, authors like Donaldson (1996) also see the tendency that with respect for local customs, Western companies become more and more familiar with foreign traditions and accept different forms of gift-giving regarding cultural expectations. A very popular and well-cited example in this context is *guanxi* as a traditional system of building social relationships (Brand and Slater 2003; Dunfee and Warren 2001; Hong 2002; Lovett et al. 1999; Millington et al. 2005; Steidlmeier 1999; Zsolnai 2007). In Chinese culture, *guanxi* is a prevalent social system which is omnipresent in daily life: in family relations as well as in dealing with public authorities, and in commercial business (Pilotta 2013). However, while some scholars link *guanxi* to corruption, anthropologists emphasize the positive effects of building trust in a highly complex business world and the possibility to overcome uncertainties in the absence of adequate bureaucratic and legal systems by a continuous reciprocal social network (Smart and Hsu 2007). In contrast to bribery, where an illegitimate or illegal transfer is granted, *guanxi* can be regarded more as a social process ensuring a long-term network of personal relationships (Steidlmeier 1999). "In this environment *guanxi* may provide a moral and ethical framework within which transactions can take place since the emphasis on trust and the maintenance of long-term relationships encourage appropriate forms of behavior within the cultural context" (Millington et al. 2005, p. 255; see also D'Souza 2003; Wong and Chan 1999).

Due to the case that in research and corporate practice the differentiation between gift-giving and bribery is often not clear and context-based, for organizations "the challenge is to determine precisely when in a particular situation a transaction constitutes a bribe and when it does not" (James 2002, p. 200). One possibility to start with might come from the macro level, i.e., the legal sphere of the economic order. What is determined in the law system as an illicit transaction has to be respected by corporate management as a major corridor for action and consequently as an explicit line between a bribe and a gift. While enforcement of transnational anti-bribery laws has been lacking for a long time in many countries, in the past years, increased efforts to prevent bribery are on the rise worldwide. Famous examples of international anti-bribery frameworks that were ratified by its members include the OECD Convention on Combating Bribery of Foreign Public Officials (1997), which focuses on the illegal act of companies and individuals paying bribes to foreign public officials, and the UN Convention against Corruption (2003) with a wider scope, including private sector corruption and bribery, both at home and abroad. However, although there exist some efforts to harmonize bribery acts on an international level the regulatory approaches are still mostly institutionalized in national law, which can vary from country to country. Therefore, in addition to the legal sphere, global businesses also resort to organizational codes of conduct to specify general rules the decision-makers should follow. The codes of conduct are furthermore relevant, as a solitary compliance with law is not sufficient for enduring business success, as we will discuss shortly. In this context, a clear ethical codex as well as an organizational culture of good conduct that explicitly states the circumstances under which gifts can be accepted as beneficial treatment, seems relevant. However, although corporate rules determine acceptable behavior on an organizational level, they are no laws in the strict sense. "Such codes may serve [more] as a 'model', in the absence of meaningful national law" (Onkvisit and Shaw 1991, p. 8). To determine the appropriate behavior for a globally operating business unit, finally also some practical wisdom (Melé 2010, 2012) seems always necessary to ensure that a granted gift does not cross the line where it could be considered as illegitimate or illegal.

After considering the difference between bribery and gift-giving, it appears interesting to explore why corporations might be motivated to bribe. Hence, some of the most relevant determinants of bribery are elaborated in the following section.

3 Why Does Bribery Appear?—Determinants of Corporate Bribery

The question relating to the factors leading to or facilitating acts of bribery in business are still vividly debated in research. Besides cultural (organizational and local) and individual aspects, most often the argument of business pressure combined with a dilemma of seemingly having to pay bribes to stay competitive in some regions is brought forth (Nichols 2012; for a review of arguments in the bribery case of Siemens see in detail Aßländer 2017). Especially latter argument will be scrutinized in the following as until now there is little research done about whether corporate bribery actually pays off or not (Zhou and Peng 2012). Regarding the positions taken in literature, there exist two competing perspectives (Cheung et al. 2012). On the one hand, bribery can be regarded as a means that might help corporations (Huntington 1968) for example "to avoid or reduce tax, to secure procurement contracts, to bypass laws and regulations, or to block the entry of potential competitors" (Wu 2005, p. 153). From this perspective, by paying bribes, corporations are sometimes able to achieve lower transaction costs.

Following these examples, on the surface, bribery could appear to be effective and efficient on the corporate level, creating monetary value. In such vein, Carmichael (1995) summarizes three central reasons, why international corporations may engage in bribery. First, in certain countries normal business transactions cannot be initiated or completed without paying bribes. Second, although a corporation may recognize bribery as morally wrong, a corporation uses bribes because "everybody does it" and it is regarded as a form of ordinary business. As a third point, Carmichael (1995) names the desperate need of a business to involve in bribery practice to ensure its short-term survival.

In addition, on an organizational level, for corporations that were once open for bribery it can be more difficult to resist future bribery activities. At first, the possibility of increased susceptibility to blackmail must be considered. For example, Kaufmann and Wei (2000) emphasize in an empirical study that companies paying bribes face more excessive bureaucracy (red tape) due to the fact that the corrupt receiving side can exert pressure by blackmailing the company or even has the power to punish a corporation for not bribing (e.g., by canceling business contracts) (Wu 2005; see also Luo 2005).

However, also cultural reasons pertaining to a development of expectations in a given organizational context must be considered, which could impede the resistance of paying bribes. Once a culture of bribery is established in corporate structures, it can be very difficult to remove this system because employees may be likely to adopt this system, accept it, or even in the worst case, are encouraged by the existing organizational bribing structures. As a very well-known and well-cited example in literature, the corruption scandal of the German company Siemens can be named (Aßländer 2017). During the 90s, Siemens established a corporate-wide bribing system, which made many employees believe that bribes were not only acceptable but also implicitly encouraged (Eberl et al. 2015; Lingnau and Fuchs 2022). When the bribery scandal finally became public, Siemens had to face various money laundering allegations and was found guilty for bribery in several countries (Wolf 2009), involving the exclusion from public contracts, an incalculable reputation damage and fines of several billion Euros. During the peak of the scandal, even a hostile takeover and a breakup were discussed as realistic scenarios for Siemens. Berghoff and Rauh (2013), two German economic historians, were even able to reveal that the Siemens bribery system, once established to acquire business contracts or to overcome bureaucratic burdens, was ineffective and unprofitable (from a business ethical perspective also see Bowie 2017). Moreover, the illicit payments made the company more vulnerable to the receiving corrupt side (e.g., for extortion). Thus, corporate culture seems to be a particularly relevant facilitator and determinant of corporate bribery. Looking at this issue from a conceptual perspective, the organizational rules and culture are generally formed by attitudes, attributions, and social factors embedded in the cultural, legal and economic background (Sanyal 2005). According to Hofstede (2001, p. 4), culture can be considered as "the collective programming of the mind". Culture is evidently a very complex, multilayered phenomenon comprising artefacts, believes and values as well as basic, unconscious assumptions (Schein and Schein 2017) and specifically relevant to prevent acts of bribery and corruption in general as it "sets the moral tone for an organization" (Luo 2005, p. 145).

In this context, it appears especially relevant to consider the particular responsibility of management as "managerial reactions set the tone by which corruption is judged in the organization" (Wated and Sanchez 2005, p. 112). Hence, the character traits prevalent in management are relevant to explore. If managers are more addicted to dark character traits like narcissism or corporate psychopathy (Lingnau et al. 2017), the risk for criminal and unethical acts is dangerously increased. Indeed, as Babiak et al. (2010, p. 175) state, "there is a dearth of empirical data on the role of psychopathy in fraud, corruption, malfeasance, and other egregious violations of the public trust. This is unfortunate, particularly if we take the view that corruption and financial disasters can be as much the fault of individuals as they are of impersonal economic and social forces". Hence, fraudulent behavior and bribery may seem for such individuals as a rational means of doing business. This risk can be increased by institutional structures like traditional incentive schemes that are at the core based on the behavioral assumptions of homo economicus, who is a prototype of a psychopath (Fuchs and Lingnau 2024). Through a person-organization fit, these kind of incentives specifically attract individuals with psychopathic traits to join an organization.

Yet, besides such extreme cases, even for non-psychopathic leaders traditional, short-term individual incentive systems can be problematic as bribery offers corrupt managers the opportunity to realize economic success on the operative level by pushing sales, profits, and short-term corporate value for their own interests, increasing the monetary compensation, while leaving the future potential risks and costs to the corporation. Concerning the costs as well as the potential short-term returns of bribery, a deeper analysis seems especially relevant. The often-voiced economic argument of bribery is analyzed in the following section, in which we explore the economic rationality of bribery payments.

4 Could It Pay to Bribe?

The economic consequences of bribery are intensely discussed in research. While some of the researchers still propose that bribery can be efficient by removing bureaucratic rigidities and "greasing the wheels of commerce" (Huntington 1968; Leff 1964; Leys 1965; Mendoza et al. 2015) other scholars focus more on the negative aspects (Fisman and Svensson 2007; Frye and Shleifer 1997; Kaufmann and Wei 2000; Luo 2005; Nichols 2012; Shleifer and Vishny 1998) and the hidden costs that bribery can cause for a corporation (Luo 2005; Wu 2005). In this sense, bribery is seen as "sand in the machine" (Ades and Di Tella 1997; Méon and Sekkat 2005) that can threaten companies' success in the short run as well as their survival in the middle or long run. Hence, bribery is linked with substantial legal risk (Wu 2005). Corporations and their managers are confronted with the law system and its enforcement. While managers who were found guilty for bribery may face fines or even jail sentences, corporations can also be confronted with immense penalties or even the nullification of existing and future contracts when illicit transactions become public.

Looking at the macroeconomic perspective, it is largely undisputed that corruption, and in specific bribery, has many negative consequences. In this context, usually the results of disproportionate prices (Kochan and Goodyear 2011), a loss of confidence in the regulatory function of state and laws (Rose-Ackerman 1997; Shah 2008), the deterrence of foreign investments (Kochan and Goodyear 2011; Nichols 2004), and the misallocation of resources (Luo 2007) as well as productivity losses are debated (Argandoña 2001; Hawthorne 2015). In such vein, the OECD (2012, p. 203) states: "Bribery and corruption are damaging to democratic institutions and the governance of corporations. They discourage investment and distort international competitive conditions. In particular, the diversion of funds through corrupt practices undermines attempts by citizens to achieve higher levels of economic, social and environmental welfare, and it impedes efforts to reduce poverty". However, from an organizational economic point of view, it might appear to be quite rational to bribe in order to obtain apparently profitable orders. In other words, as long as the expected profit from an order using bribery is greater than the expected value of profit without bribery, it might appear to be rational to bribe for obtaining a desired order.

Taking a mathematical point of view (Lingnau and Kokot 2015, also see Klein 2016; Becker 1968), this situation can be modelled as follows:

- $p(B^-)$: Probability of winning the contract without bribery
- $p(B^+)$: Probability of winning the contract with bribery
- π^-: Profit from contract without bribery
- π^+: Profit from contract with bribery
- p(conviction): Probability that the bribe will be detected
- C(conviction): Costs of bribery detection (e.g., state sanctions)

From these variables, using a purely economic approach, it appears to be profitable to bribe when the following condition is fulfilled:

If $p(B^-) * \pi^- < p(B^+) * \pi^+ - p(conviction) * C(conviction)$

then bribery is rational from a traditional economic perspective. Yet, such economic thinking must be seriously questioned. While under conditions of decision-making under certainty or risk, where probabilities are known (Angner 2016; Baron 2008; Daft 2016) such an approach might be conceivable, most decision-making takes place under conditions of more or less uncertainty, where the probabilities of events are unknown. In this context, as several corporate scandals from the past show, most often the probability of disclosure as well as the costs incurred have been underestimated. As real-life decisions often imply solely subjectively estimated probabilities, being a decision under a larger degree of uncertainty, many bribing actors have been wrongly assuming that a probability of detection would be close to zero (i.e., *p(conviction)* $\rightarrow$ 0). However, due to increasingly connected, critically observant stakeholders and the number of people involved in business, the probability that eventually the act of bribery will become public rather tends to converge to one (i.e., *p(conviction)* $\rightarrow$ 1) in an often rather short period of time (see mathematically also Grimes 2016). Such is similarly emphasized by Eicher (2009, p. 1) stating that "[p]ublic awareness has risen through the efforts of NGOs, international agencies, and scandals in the media". In addition, various regulatory efforts, such as the US Federal Sentencing Guidelines for Organizations, have increased the cost of bribery when uncovered. Lastly, also the expected profits due to bribery are in practice often misjudged since the incidental bribery costs are often underestimated.

In summary, empirical findings show that the deliberate disregard of legal regulations to increase current economic success may seem rational in the short-term, but also entails considerable risks in the long-term. As Luo (2005, p. 138) states: "Although it is possible that returns from corruption may outweigh costs incurred for a specific deal in the short term, it is unlikely, if not impossible, that a firm can build on corruption to achieve [...] abnormal profitability in the long run". For business, this obviously means that there is a critical need to take care of consistent integration and standardization within corporate borders. This can take the form of compliance management, for example, which aims to prevent criminal behavior within the enterprise. When implementing a compliance management system, the focus is on detailed regulations as a guideline and the fear of punishment as a motivational factor. Such is the reason why compliance is often considered as part of a wider corporate risk management.

As costs and detection probability are related to public institutions, also state regulation represents a major factor in preventing the occurrence of bribery. As emphasized, decision-makers often justify acts of bribery by arguing with the dilemma situation they face. Such dilemma can be analyzed in the traditional game theoretical model of the prisoner's dilemma where it seems individually profitable to bribe, irrespective of what the competitors do (Kochan and Goodyear 2011). If everyone bribes, this leads to an inferior situation for all the actors involved (as well as the society as a whole), calling for a change in the expected payoffs to resolve the prisoner's dilemma concerning the wrong incentives as well as mitigating the negative outcomes for all actors involved. Such game theoretical analysis is also prominent among some business ethicists, namely in the school of German business ethicist Karl Homann, arguing for better regulation to overcome those dilemma situations (e.g., Homann 2016; Homann and Blome-Drees 1992; Homann and Lütge 2013). In such vein, states could increase the frequency of checks and the level of fines to make bribery less attractive.

While at first it might appear to be sensible to argue with such game theoretical dilemma structure, calling for increased regulation, the argument exhibits at least two shortcomings. At first, as already discussed, it seems evident that the prisoner's dilemma focusses from the perspective of incentives mainly on the individually apparently profitable short-term yields due to bribery. If one considers the long-term risk involved for the deciding business unit, it seems rather questionable that it may *always* be the superior strategy to bribe for the actors, no matter what the competitors do. Furthermore, a sole focus on better regulation, i.e., a focus on the macro level of improving the legal framework to make bribery less attractive by imposing larger fines, creating stricter control institutions, etc., would face an extreme limitation on the global level. As Crane et al. (2014, p. 140) emphasize, globally active "corporations operate in a broad variety of geopolitical contexts where governments are unable or unwilling to regulate them effectively". Since there is no global state and government existing (Pies et al. 2010) that might be credible to impose such fines, it seems implausible to assume that there would be a possibility for the near future to regulate the issue of bribery by means of law alone. In such vein, likewise Wettstein and Waddock (2005, p. 304) emphasize: "The ongoing economic globalization, however, has rendered this state-centered view increasingly inadequate." In addition, a more than elementary trust in the solitary regulatory power of law always seems to impose the danger of overestimating the capabilities of formalized state norms since law represents merely the basic societal norms to adhere to (Crane and Matten 2016).

Such however, as we will delineate shortly, seems not seriously necessary from a business ethical point of view. Up to now, the question of why corporations should deal with ethical questions from a business point of view has been argued from a purely economic perspective asserting that the failure to take the long-term consequences of bribery into account can lead to "ethical costs" that may exceed the savings sought by such behavior by orders of magnitude. In order to avoid such ethical costs, ethical aspects must be integrated into economic decisions. Ultimately, however, this argumentation (e.g., in the case of Homann 2002) is only about a continuation of economics by other means because ethical aspects are considered solely under an economic benefit calculation (Ulrich 2008).

As the following considerations show, such a narrow reflection on the issue of bribery is not enough. Although cost–benefit calculations are certainly a predominant concept in managerial decision-making, a sole focus on economic terms would be too limited from a business ethical perspective. As we discuss in the following, it seems relevant to transcend simple economic calculations and consider some normative aspects of bribery in business.

5 Limits of the Traditional Approach and Some Business Ethical Reflections

One first approach to critically scrutinize the economic argumentation outlined above might start from a traditional normative, e.g., Kantian perspective. From such, Bowie (1999, p. 14) argues that "if a maxim permitting bribery were universalized, it could not pass the test of the categorical imperative [...]. If everybody offered bribes, then the practice of trying to make secret attempts to gain special advantages would make no sense" (also see Bowie 2017, p. 172). Although ethically sound, for practical reasons, usually such purely philosophical arguments run into danger of being not persuasive enough for practitioners, claiming that ethical conduct has to be not solely intellectually convincing and ethically reasonable, but also in some sense valuable for business and feasible in helping them to conduct their business in a successful manner. Business ethics as an applied ethics is therefore confronted with the task of putting "formal ethical considerations developed a priori into practical use in real-life situations" (Zimmerli and Aßländer 2010, p. 37). Thus, an approach seems necessary that combines a business-intrinsic relevance for ethics without conversely economizing business ethics, leading to the danger of committing a category mistake (Fuchs 2022; Lingnau and Fuchs 2022; Ulrich 2008).

A fruitful approach that transcends the previous economic considerations, going beyond pure cost–benefit calculations, stresses the danger that businesses lose their "license to operate" (for this concept see Morrison 2014; Reinhardt 2005; Thomson and Boutilier 2011; Wilburn and Wilburn 2011), i.e., their legitimacy, if they do not respect legitimate stakeholder claims, for example by not paying attention to social implications of bribery. Such concept seems especially relevant, as several business scandals demonstrate that the legitimacy of a business entity might be seriously endangered if corruption processes become public. From this perspective, legitimacy appears to be a major prerequisite for permanent business success, which transcends pure cost–benefit calculations as it is not sensibly measureable in terms of monetary units. On the contrary, the criteria to be applied here is largely not subject to discounting at all. Therefore, the question whether an action "pays off" in an *ethical* perspective represents a category error, since legitimacy (similar to related concepts like trust or reputation) and (discounted) cash flows obviously belong to different categories. Hence, the question by how many monetary units the legitimacy will increase due to an improved ethical conduct is obviously absurd and cannot be answered.

The concept of legitimacy itself has stirred an intense debate (for an overview see Bitektine 2011; Deephouse et al. 2017; Suddaby et al. 2017). One classical and often cited definition by Suchman (1995, p. 574) perceives legitimacy as "a generalized perception or assumption that the actions of an entity are desirable, proper, or appropriate within some socially constructed system of norms, values, beliefs, and definitions". Also some newer definitions follow such an approach. For instance, Deephouse and Carter (2005, p. 332) define "legitimacy as the social acceptance resulting from adherence to regulative, normative or cognitive norms and expectations". One point of criticism that could be voiced with these definitions is that the concept of such "social legitimacy" might be too culturally dependent (ethical relativism), calling e.g., for an independent normative-ethically justified legitimacy instead (e.g., Melé and Armengou 2016) that does not solely build upon an analysis of existing legitimacy expectations for managerial decision support but rather intends to evaluate business decisions by a defined basic normative argument, like the human rights discourse or virtue ethics. From a philosophical standpoint, such argument seems certainly relevant. Yet, an approach like this could run into danger of being considered less persuasive for practical decision purposes. To avoid such, it seems fruitful to differentiate two different levels of legitimacy analysis (Fuchs 2022).

From a perspective of providing decision-support, it seems clear that *every* business must, to a certain degree, adapt to culturally specific perceptions of legitimacy in the given area it operates—*whatever* theses expectations might be in concrete terms. Otherwise, the business will run into danger of facing a withdrawal of life-critical resources. For corporations to survive, it appears paramount to keep their legitimacy, because "organizations cannot run their operations unless the communities in which they operate accept their presence" (Corvellec 2007, pp. 138–139). Such view is supported by Carroll et al. (2012, p. 1) underlining that "[i]nstitutions cannot survive and prosper without the acceptance, endorsement and support of others in the society" (also see Deephouse et al. 2017, p. 27). Such support however might be seriously endangered when corruption practices of bribery become evident, as visible in recent corporate scandals: "In extreme cases, a lack of legitimacy can jeopardize [...] even the survival of a company" (Melé and Armengou 2016, p. 730).

In a second step, however, one could very well debate whether the legitimacy perceptions of a given society are from some established more "objective" moral point of view criticizable, such that a business should rather perish than adapt to a potential "culture of devils". In such vein, it seems useful to differentiate between the necessary and sufficient condition of legitimacy (Fuchs 2022). To sustain a business unit, it would be empirically first of all *necessary* to adequately adapt (at least in the short-term) to specific moral expectations of a community. However, from a reflective ethical perspective these norms are themselves again subject to a critical analysis to come to an *ethically sufficient* legitimacy. This critical reflection can be based on a descriptive ethics like commonly shared basic values, or on a normative-ethical approach, which is based on specific normative assumptions and theories (see Fig. 1).

As a note and with regard to a critique of seemingly accepting a "culture of devils" in case of merely applying a descriptive perspective, it seems doubtful whether such a truly unethical society, concerning the general population's legitimacy perceptions,

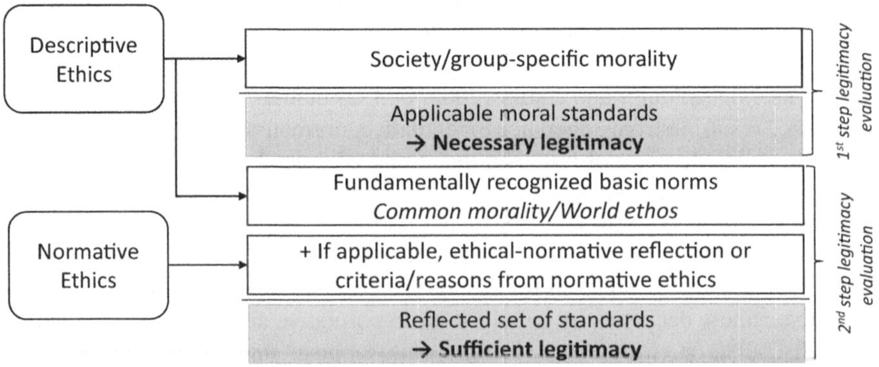

Fig. 1 Two-step evaluation of legitimacy (based on Fuchs 2022)

would actually be that likely to exist. As empirical research, for instance on a world ethos (Küng 1998, 2004), demonstrates, the basic norms usually differ not that much throughout all societies (also see Arnold et al. 2013; Melé 2009). Lying and cheating are for instance in no known society openly supported. Consequently, bribery is nearly always seen quite critically and most of the argument (as outlined above) focuses on the issue of when gifts are overstepping the line of becoming an act of bribery. In such vein, Cragg (2018, p. 577) emphasizes that although "bribery is common in many parts of the world today [...] bribery is universally condemned. If this were not the case, it would not be necessary to conceal bribes behind a veil of secrecy". Similarly, Noonan (1987, p. 702) emphasizes that "[b]ribes are universally shameful". Finally, also in a broader perspective, Sandholtz and Koetzle (2000, p. 34) summarize: "Corrupt acts are, in every definition, improper or illegitimate". Hence, it seems very doubtful that even when acts of corruption are not that actively pursued by public authorities, the general population is truly *embracing* the corrupted system of, for instance, regular bribery payments. Rather, it appears to be that acts of bribery are most often de-facto accepted, as an individual is unable to change the institutionalized system of power and influence in a given society. Thus, a resigned or factual acceptance of bribery does not necessarily imply that such acts are widely considered as certainly legitimate.

Conversely, from a managerial point of view, it appears highly relevant to ensure that an overall higher level of support is achieved, providing a business with some buffer of legitimization toward its operations (Fuchs 2022). In such vein, it seems of great importance to consider the significance of a systematic legitimacy management, which we discuss in the following.

6 The Relevance of Legitimacy Management

In the previous section, we emphasized the relevance of keeping the societally granted license to operate from a managerial perspective to safeguard a business's existence in the long-run. However, in this context, the question arises if a strict reliance on legal standards (compliance management) is already sufficient to provide enough business legitimacy as part of a larger legitimacy management (e.g., Lenssen and Vorobey 2005). Looking at some current empirical studies, it becomes evident that the acceptance of the market economy system and its underlying economic principle has reached historic lows (Horváth and Barton 2016). If the economic order of a democratic society were (almost) perfect, any legal action would also be legitimate. However, as the economic order evidently is rather imperfect, exclusively legal behavior is usually insufficient to justify the legitimacy of actions, constituting a case for "beyond compliance". In other terms: "Compliance is not enough"! (Kennedy-Glans and Schulz 2005, p. xix; Lingnau and Fuchs 2022, p. 1033; Wheeler 1992, p. 10). This perspective is also widely shared in practice. For instance, Caterpillar Tractor Company's former Worldwide Code of Conduct explicitly stated: "The law is a floor. Ethical business conduct should normally exist at a level well above the minimum required by law" (Caterpillar 1985, p. 112). However, although this statement cannot be found in Caterpillar's latest approved Code of Conduct, the argumentation did not change: "While we conduct our business within the framework of applicable laws and regulations, for us, compliance with the law is not enough" (Caterpillar 2023, p. 3). Consequently, as compliance is not sufficient, substantiated moral issues require some consideration on the organizational level to ensure that the corporation keeps its societally granted license to operate, ensuring its long-term survival: "Corporations, if they are to remain in business for the long term, must operate within a context of community values" (Burnup 1993, p. 172).

As the press and the critical public clearly no longer automatically accept legal behavior as legitimate, the responsibility of legitimizing one's business has returned to the enterprise and the individuals working for it. Even the smallest transgressions of norms, which might appear to be efficient in the short-term, can have life-critical consequences for a corporation on a normative level. Especially in times of WikiLeaks, and other related web platforms, it can be increasingly expected that a corrupt behavior in the middle or long run will not remain undetected (Grimes 2016). Consequently, executives are increasingly forced to justify their behavior in an argumentative and ethical manner and in specific the societal impact of their actions. However, as evident, many executives still struggle to provide sound ethical answers to justify their business conduct toward a critical stakeholder audience. This seems essential, as it is the task of management to integrate the partially conflicting demands of numerous stakeholder groups in order to effectively ensure the survival of the business (Lingnau 2009). Adding to such considerations, Rüegg-Stürm (2005, p. 22) emphasizes that "[i]n the case of conflicting claims of the stakeholder groups, *respectful consideration and carefully reasoned evaluation*" is necessary, and that "the explanation for the final decision must *be understood* [.]", calling for a necessity of constant legitimization of businesses.

Hence, an inappropriate (unethical) discrimination against a stakeholder group like paying a bribe to illegitimately win a contract carries the risk of losing legitimacy. In such vein, it is the management's task to integrate the claims of the numerous stakeholder groups in order to ensure the legitimacy of entrepreneurial activity. Such thoughts can for example be connected to the considerations by Rüegg-Stürm (2005), differentiating between normative, strategic, and operative management. While at the normative level, business justification and legitimacy is situated, the strategic level focusses on generating and maintaining success potentials, whilst finally at the operative level the generation of success (by the utilization of success potentials) is located. Being the corporation's figurehead and final decision-authority, top-management has the duty of safeguarding the legitimacy of business, which therefore can be characterized as a genuine management task. The three levels of management are from an ethical perspective inseparably linked with each other. Hence, the strategic goals of creating and sustaining success potentials must not contradict the normative level of management in pursuit of saving the legitimacy of the organization, and operative processes must in turn not conflict with the strategic corporate goals. Such is also emphasized by Rüegg-Stürm (2005, pp. 22–23), asserting: "[I]n terms of normative management, strategic and operative decision-making processes should be firmly based upon ethical considerations". Therefore, also in the context of bribery, the much broader negative consequences are evident as such conduct may have a detrimental impact on the credibility and legitimacy of business as the highlighted business scandals exhibit. As Kayes et al. (2007, p. 61) state: "Ethical lapses by employees can put organizations at substantial risk. Although improved compliance procedures can help limit this risk, successful efforts must extend beyond compliance [...]." Therefore, to sustain the continuous existence of the corporation it seems vital to transcend simple cost–benefit calculations as well as *solely* legalistic interpretations of good conduct to avoid acts of bribery, but rather focus on a corporate culture that internalizes also the motives for legitimacy to sustain the long-term existence of the business. In such vein, also Harisalo and Stenvall (2004, p. 62) emphasize: "In order to earn and maintain legitimacy, management must not only be legally right, they must also be morally right". As stakeholders are increasingly questioning corporate practices, merely trying to manage stakeholder perceptions (e.g., Lawrence et al. 1997; Murphy and Murphy 2004) is not enough. Rather, legitimate expectations have to be truly considered and management has to be able to answer these questions in a constructive discourse with the organization's stakeholders (Pies et al. 2010). In such way, legitimacy management represents an essential extension of traditional approaches to prevent harm by sole legal adherence. It rather constitutes a much more proactive way of engaging with stakeholders to ensure that business always stays within the limits of moral permissibility. In this context, it seems particularly fruitful to enable a corporate culture of integrity (Holder 2016; Paine 1994; Vandekerckhove 2016) representing one major building block for such systematic legitimacy management. Finally, also for management there are some clear implications of the aforementioned considerations. Management itself must lead with a good example and be open for serious ethical discussions. As individuals or even smaller groups of individuals are rarely able to consider all the relevant normative details, it seems especially beneficial to take into account a diverse set of perspectives. Such could indeed then enable management to play an active and constructive role in the discourse and help it better understand when providing a gift seems a necessary kindness to introduce and maintain business relations and when it runs into danger of overstepping the thin line of initiating an illegitimately preferred treatment. These considerations are summarized in the following Table 1.

7 Summary and Conclusions

In this chapter, we provided a critical reflection on the consequences of corporate bribery. At first, we discussed the notion itself, differentiating it from legitimate forms of gift-giving. Although a clear separation between those two forms of exchange is often hard to make, central features of bribery, like the illegitimate intention to exert influence on a decision, might help to identify such acts. After these initial considerations, we discussed some prominent arguments as to why companies are inclined to pay bribes. As became evident, besides organizational-cultural or individual reasons,

Table 1 Major differences between compliance and legitimacy management

	Compliancemanagement	Legitimacy management
Basic norms	Legal requirements	Moral expectations of society, ethical standards
Motivation	Conformity to minimal legal standards, specifically with laws, to avoid state sanctions	Protecting societal license to operate, long-term safeguarding and fostering business development capabilities
Control logic	More reactive	More proactive
Organizational culture	Focus on control measures, limiting discretionary room of action in order to prevent illegal activity	Culture of integrity, enabling moral action and "self-control" of the employee, supportive control measures
Primary addressee of communication	Dominantly: legal authorities (external)	Stakeholders (internal and external)
Managerial implications	Supervision and control in line with legal requirements	Leading with good example, justification of actions toward stakeholders, proactive engagement with stakeholders

one major reason brought forth is often of economic nature. In this context, it is frequently argued that it might pay to bribe, or conversely, that non-bribery would be economically inefficient. To analyze such argument, a traditional economic analysis was provided, leading to the conclusion that the apparently positive results of bribery are often questionable, which results from underestimating the probability of detection as well as the costs inferred when the corrupt act is detected.

However, as we emphasized, such a purely economic approach reaches far too short since it solitarily remains on the operative level of business, neglecting the strategic and, importantly, the normative implications of bribery. Hence, not only economic costs and benefits of bribery are to be considered, but also the much longer lasting and often unquantifiable outcomes, resulting from the degradation of publicly perceived legitimacy. Such seems important to highlight from a perspective of business ethics, because legitimacy cannot be seriously quantified in monetary terms. As we emphasized, a lack of legitimacy can have detrimental impacts on the existence of businesses when investors withdraw their financial support, well-educated employees quit the organization or customers are unwilling to purchase corporate goods or services. Thus, it seems important to consider the wide spread negative impact of illegitimate acts. Luo (2005, pp. 138–139) rightly emphasizes in this context that "[t]he image effect in the eye of the public is enduring, far-reaching and more significant than the financial figures [...]." Such is especially true for Western businesses that are confronted with an increasingly critical public opinion as well as the ability of key stakeholders to look for alternatives if an organization loses its legitimacy. As a compliance approach is not enough for corporate legitimacy, it appears especially relevant to focus on a legitimacy management in which gift-giving is certainly recognized as vital prerequisite for doing business in certain cultures, for instance in many Asian countries (e.g., D'Souza 2003), but clearly refrains from encouraging any active engagement to bribe for gaining an illegitimately preferred treatment in business contracts.

From a business ethical perspective, we think that finally more tools for the normative level of management are urgently required, especially as many enterprises operate internationally in a diverse set of stakeholder expectations: "Today companies do not operate within national boundaries—where the rules are straight-forward and understood—but in multiple countries, each with its own set of norms and rules" (Eicher 2009, p. 1). Hence, it seems fruitful to provide management not solely with an overview, as most often in form of reports, on the legal sphere of their operations, but likewise to sensitize the employees for ethical issues and to give them the tools for sound ethical reasoning. With such tools, also management could better see which norms might probably conflict and which gifts can be granted without overstepping legitimate expectations by other stakeholders.

References

Ades, A., & Di Tella, R. (1997). The new economics of corruption: A survey and some new results. *Political Studies, 45*(3), 496–515.

Angner, E. (2016). *A course in behavioral economics* (2nd ed.). Basingstoke: Palgrave Macmillan.

Argandoña, A. (2001). Corruption: The corporate perspective. *Business Ethics: A European Review, 10*(2), 163–175.

Argandoña, A. (2005). Corruption and companies: The use of facilitating payments. *Journal of Business Ethics, 60*(3), 251–264.

Arnold, D. G., Beauchamp, T. L., & Bowie, N. E. (2013). *Ethical theory and business* (9th ed.). Upper Saddle River, NJ: Pearson.

Aßländer, M. S. (2017). Corruption in the IT branch – The example of Siemens. In M. S. Aßländer & S. Hudson (Eds.), *The handbook of business and corruption: Cross-sectoral experiences* (pp. 209–236). Bingley: Emerald Publishing.

Babiak, P., Neumann, C. S., & Hare, R. D. (2010). Corporate psychopathy: Talking the walk. *Behavioral Sciences & the Law, 28*(2), 174–193.

Baron, J. (2008). *Thinking and deciding* (4th ed.). Cambridge: Cambridge University Press.

Beck, P. J., & Maher, M. W. (1986). A comparison of bribery and bidding in thin markets. *Economics Letters, 20*(1), 1–5.

Becker, G. S. (1968). Crime and punishment: An economic approach. *The Journal of Political Economy, 76*(2), 169–217.

Berghoff, H., & Rauh, C. (2013). *Korruption rechnet sich nicht*. http://www.faz.net/aktuell/wirtschaft/unternehmen/bestechende-grossunternehmen-korruptionrechnet-sich-nicht-12050962.html. Accessed 30 July 2024.

Bitektine, A. (2011). Toward a theory of social judgments of organizations: The case of legitimacy, reputation, and status. *Academy of Management Review, 36*(1), 151–179.

Bowie, N. E. (1999). A Kantian approach to business ethics. In R. E. Frederick (Ed.), *A companion to business ethics* (pp. 3–16). Malden, MA: Blackwell Publishers.

Bowie, N. E. (2017). *Business ethics: A Kantian perspective* (2nd ed.). New York: Cambridge University Press.

Brand, V., & Slater, A. (2003). Using a qualitative approach to gain insights into the business ethics experiences of Australian managers in China. *Journal of Business Ethics, 45*(3), 167–182.

Burke, R. J. (2016). Crime and corruption in organizations. In R. J. Burke, E. C. Tomlinson & C. L. Cooper (Eds.), *Crime and corruption in organizations: Why it occurs and what to do about it* (pp. 3–65). London: Routledge.

Burnup, C. (1993). "Beyond compliance": A corporate response to environmental ethics. In C. A. J. Coady & C. J. G. Sampford (Eds.), *Business, ethics, and the law* (pp. 172–186). Sydney: Federation Press.

Carmichael, S. (1995). *Business ethics: The new bottom line*. London: Demos.

Carroll, A. B., Lipartito, K. J., Post, J. E., & Werhane, P. H. (2012). *Corporate responsibility: The American experience*. Cambridge: Cambridge University Press.

Caterpillar (1985). Excerpts from Caterpillar tractor company's code of worldwide business conduct and operationg principles. In G. R. Laczniak & P. E. Murphy (Eds.), *Marketing ethics: Guidelines for managers* (pp. 111–116). Lexington: Lexington Books.

Caterpillar (2023). *Our values in action: Caterpillar's code of conduct*. http://s7d2.scene7.com/is/content/Caterpillar/CM20190326-783ea-8cfad. Accessed 30 July 2024.

Cheung, Y. L., Rau, P. R., & Stouraitis, A. (2012). *How much do firms pay as bribes and what benefits do they get? Evidence from corruption cases worldwide*. NBER Working Paper No. 17981. http://www.nber.org/papers/w17981. 30 October 2018.

Cleveland, M., Favo, C. M., Frecka, T. J., & Owens, C. L. (2009). Trends in the international fight against bribery and corruption. *Journal of Business Ethics, 90*(2), 199–244.

Corvellec, H. (2007). Arguing for a license to operate: The case of the Swedish wind power industry. *Corporate Communications: An International Journal, 12*(2), 129–144.

Cragg, W. (2018). Corruption, bribery, and moral norms across national boundaries. In E. Heath, B. Kaldis & A. Marcoux (Eds.), *The Routledge companion to business ethics* (pp. 573–589). London: Routledge.

Crane, A., & Matten, D. (2016). *Business ethics: Managing corporate citizenship and sustainability in the age of globalization* (4th ed.). Oxford: Oxford University Press.

Crane, A., Palazzo, G., Spence, L. J., & Matten, D. (2014). Contesting the value of "creating shared value". *California Management Review, 56*(2), 130–153.

Daft, R. L. (2016). *Management* (12th ed.). Boston, MA: South-Western Cengage Learning.

Davies, G., Whelan, S., Foley, A., & Walsh, M. (2010). Gifts and gifting. *International Journal of Management Reviews, 12*(4), 413–434.

Deephouse, D. L., & Carter, S. M. (2005). An examination of differences between organizational legitimacy and organizational reputation. *Journal of Management Studies, 42*(2), 329–360.

Deephouse, D. L., Bundy, J., Tost, L. P., & Suchman, M. C. (2017). Organizational legitimacy: Six key questions. In R. Greenwood, C. Oliver, T. B. Lawrence & R. E. Meyer (Eds.), *The SAGE handbook of organizational institutionalism* (2nd ed.) (pp. 27–54). London: Sage Publications.

Donaldson, T. (1996). Values in tension: Ethics away from home. *Harvard Business Review, 74*(5), 48–62.

Dorsch, M. J., & Kelley, S. W. (1994). An investigation into the intentions of purchasing executives to reciprocate vendor gifts. *Journal of the Academy of Marketing Science, 22*(4), 315–327.

Douglas, M. (2002). Foreword: No free gifts. In M. Mauss (Ed.), *The gift: The form and reason for exchange in archaic societies*. London: Routledge.

D'Souza, C. (2003). An inference of gift-giving within Asian business culture. *Asia Pacific Journal of Marketing and Logistics, 15*(1/2), 27–38.

Dunfee, T. W., & Warren, D. E. (2001). Is guanxi ethical? A normative analysis of doing business in China. *Journal of Business Ethics*, 32(3), 191–204.

Eberl, P., Geiger, D., & Aßländer, M. S. (2015). Repairing trust in an organization after integrity violations: The ambivalence of organizational rule adjustments. *Organization Studies, 36*(9), 1205–1235.

Eicher, S. (2009). *Corruption in international business: The challenge of cultural and legal diversity*. Farnha: Routledge.

Fisher, C. M., Lovell, A., & Valero-Silva, N. (2013). *Business ethics and values* (4th ed.). Harlow: Pearson.

Fisman, R., & Svensson, J. (2007). Are corruption and taxation really harmful to growth? Firm level evidence. *Journal of Development Economics, 83*(1), 63–75.
Fritzsche, D. J. (2005). *Business ethics: A global managerial perspective.* New York: Mason Crest.
Frye, T., & Shleifer, A. (1997). The invisible hand and the grabbing hand. *American Economic Review, 87*(2), 354–358.
Fuchs, F. (2022). *Subjektunabhängige, analytische Unternehmensethik: Begründung und Relevanz als praktisch-normative Betriebswirtschaftslehre.* Wiesbaden: Springer.
Fuchs, F., & Lingnau, V. (2024). The homo economicus as a prototype of a psychopath? A conceptual analysis and implications for business research and teaching. *Journal of Business Ethics, 195*, 763–777.
Gordon, K., & Miyake, M. (2001). Business approaches to combating bribery: A study of codes of conduct. *Journal of Business Ethics, 34*(3–4), 161–173.
Gottschalk, P. (2016). *Investigation and prevention of financial crime: Knowledge management, intelligence strategy and executive leadership.* London, New York: Routledge.
Graycar, A., & Jancsics, D. (2017). Gift giving and corruption. *International Journal of Public Administration, 40*(12), 1013–1023.
Grimes, D. R. (2016). On the viability of conspiratorial beliefs. *PloS one, 11*(1), 1–17.
Harisalo, R., & Stenvall, J. (2004). Trust as capital: The foundation of management. In M.-L. Huotari & M. Iivonen (Eds.), *Trust in knowledge management and systems in organizations* (pp. 51–81). Hershey, PA, London: IGI Global.
Hawthorne, O. E. (2015). *Do international corruption metrics matter? The impact of Transparency International's corruption perception index.* Lanham, MD: Lexington Books.
Healy, P. M., & Serafeim, G. (2016). An analysis of firms' self-reported anticorruption efforts. *The Accounting Review, 91*(2), 489–511.
Hofstede, G. (2001). *Culture's consequences: Comparing values, behaviors, institutions and organizations across nations* (2nd ed.). Thousand Oaks: Sage Publications.
Holder, F. (2016). *Integrity in business: Developing ethical behavior across cultures and jurisdictions.* Abingdon, New York: Routledge.
Holmes, L. (2015). *Corruption: A very short introduction.* Oxford: Oxford University Press.
Homann, K. (2002). Ökonomik: Fortsetzung der Ethik mit anderen Mitteln. In C. Lütge (Ed.), *Vorteile und Anreize: Zur Grundlegung einer Ethik der Zukunft* (pp. 243–266). Tübingen: Mohr Siebeck.
Homann, K. (2016). Theory strategies of business ethics. In C. Lütge & N. Mukerji (Eds.), *Order ethics: An ethical framework for the social market economy* (pp. 37–54). Munich: Springer.
Homann, K., & Blome-Drees, F. (1992). *Wirtschafts- und Unternehmensethik.* Göttingen: Vandenhoeck & Ruprecht.
Homann, K., & Lütge, C. (2013). *Einführung in die Wirtschaftsethik* (3rd ed.). Münster: LIT.
Hong, Y. (2002). Business ethics: Is it useful? An empirical study of Chinese enterprise. *Business Ethics: A European Review, 11*(4), 335–342.
Horváth, D., & Barton, D. (2016). Conclusion: Capitalism re-imagined. In D. Barton, D. Horváth & M. Kipping (Eds.), *Re-imagining capitalism* (pp. 319–331). Oxford: Oxford University Press.
Huntington, S. P. (1968). *Political order in changing societies.* New Haven: Yale University Press.
James, H. S. (2002). When is a bribe a bribe? Teaching a workable definition of bribery. *Teaching Business Ethics, 6*(2), 199–217.
Kaufmann, D., & Wei, S.-J. (2000). *Does "grease money" speed up the wheels of commerce?* IMF Working paper No. 00/64. https://www.imf.org/external/pubs/ft/wp/2000/wp0064.pdf. Accessed 30 July 2024.
Kayes, D. C., Stirling, D., & Nielsen, T. M. (2007). Building organizational integrity. *Business Horizons, 50*(1), 61–70.
Kennedy-Glans, D., & Schulz, R. (2005). *Corporate integrity: A toolkit for managing beyond compliance.* Mississauga: Wiley.
Klein, G. (2016). *Ethics in accounting: A decision-making approach*, Hoboken, NJ: Wiley & Sons.

Kochan, N., & Goodyear, R. (2011). *Corruption: The new corporate challenge*. Basingstoke: Palgrave Macmillan.
Küng, H. (1998). *A global ethic for global politics and economics*. Oxford: Oxford University Press.
Küng, H. (2004). *Global responsibility: In search of a new world ethic*. Eugene, OR: Wipf and Stock.
Lawrence, T. B., Wickins, D., & Phillips, N. (1997). Managing legitimacy in ecotourism. *Tourism Management, 18*(5), 307–316.
Leff, N. H. (1964). Economic development through bureaucratic corruption. *American Behavioral Scientist, 8*(3), 8–14.
Lenssen, G., & Vorobey, V. (2005). The role of business in society in Europe. In A. Habisch, J. Jonker, M. Wegner & R. Schmidpeter (Eds.), *Corporate social responsibility across Europe* (pp. 357–375). Berlin, Heidelberg: Springer.
Leys, C. (1965). What is the problem about corruption? *The Journal of Modern African Studies, 3*(2), 215–230.
Lingnau, V. (2009). Shareholder value as the basis for a 'controlling' theory? A knowledge based approach with analogies to marketing science. *Management, 13*(spec. ed.), 25–46.
Lingnau, V., & Fuchs, F. (2022). Controlling und Ethik. In W. Becker & P. Ulrich (Eds.), *Handbuch Controlling* (2nd ed.) (pp. 1023–1041). Wiesbaden: Springer.
Lingnau, V., & Kokot, K. (2015). Rationalprinzip, Illegalität, Legalität und Legitimität. In K. Vorbohle, J. H. Quandt & C. Schank (Eds.), *Verantwortung in der globalen Wertschöpfung* (pp. 27–48). Munich: Rainer Hampp.
Lingnau, V., Fuchs, F., & Dehne-Niemann, T. E. (2017). The influence of psychopathic traits on the acceptance of white-collar crime: do corporate psychopaths cook the books and misuse the news? *Journal of Business Economics, 87*(9), 1193–1227.
Lovett, S., Simmons, L. C., & Kali, R. (1999). Guanxi versus the market: Ethics and efficiency. *Journal of International Business Studies, 30*(2), 231–247.
Lui, F. T. (1985). An equilibrium queuing model of bribery. *Journal of Political Economy, 93*(4), 760–781.
Luo, Y. (2005). An organizational perspective of corruption. *Management and Organization Review, 1*(1), 119–154.
Luo, Y. (2007). *Guanxi and business* (2nd ed.). Hackensack, London: World Scientific.
Melé, D. (2009). *Business ethics in action: Seeking human excellence in organizations*. Basingstoke: Palgrave Macmillan.
Melé, D. (2010). Practical wisdom in managerial decision making. *Journal of Management Development, 29*(7/8), 637–645.
Melé, D. (2012). *Management ethics: Placing ethics at the core of good management*. Basingstoke, New York: Palgrave Macmillan.
Melé, D., & Armengou, J. (2016). Moral legitimacy in controversial projects and its relationship with social license to operate: A case study. *Journal of Business Ethics, 136*(4), 729–742.
Mendoza, R. U., Lim, R. A., & Ong Lopez, A. (2015). Grease or sand in the wheels of commerce? Firm level evidence on corruption and SMES. *Journal of International Development, 27*(4), 415–439.
Méon, P. G., & Sekkat, K. (2005). Does corruption grease or sand the wheels of growth? *Public Choice, 122*(1–2), 69–97.
Millington, A., Eberhardt, M., & Wilkinson, B. (2005). Gift giving, guanxi and illicit payments in buyer–supplier relations in China: Analysing the experience of UK companies. *Journal of Business Ethics, 57*(3), 255–268.
Morrison, J. (2014). *The social license: How to keep your organization legitimate*. Basingstoke: John Morrison.
Mousavi, P., & Pourkiani, M. (2013). Administrative corruption: Ways of tackling the problem. *European Online Journal of Natural and Social Sciences, 2*(3), 178–187.
Murphy, P. E., & Murphy, A. E. (2004). *Strategic management for tourism communities: Bridging the gaps*. Clevedon et al.: Channel View Publications.

Nichols, P. M. (2004). Corruption as an assurance problem. *American University International Law Review, 6*(19), 1307–1349.

Nichols, P. M. (2012). The business case for complying with bribery laws. *American Business Law Journal, 49*(2), 325–368.

Noonan, J. T. (1987). *Bribes: The intellectual history of a moral idea.* Berkeley: University of Carlifornia Press.

OECD (1997). *Convention on combating bribery of foreign public officials in international business transactions.* https://legalinstruments.oecd.org/en/instruments/OECD-LEGAL-0293. Accessed 30 July 2024.

OECD (2012). *Annual report on the OECD guidelines for multinational enterprises 2012: Mediation and consensus building.* Paris: OECD Publishing.

OECD (2013). *Bribery and corruption awareness handbook for tax examiners and tax auditors.* https://read.oecd-ilibrary.org/taxation/bribery-and-corruption-awareness-handbook-for-tax-examiners-and-tax-auditors_9789264205376-en#page3. Accessed 30 July 2024.

Onkvisit, S., & Shaw, J. J. (1991). International corporate bribery: Some legal, cultural, economic, ethical-philosophical, and marketing considerations. *Journal of Global Marketing, 4*(2), 5–20.

Paine, L. S. (1994). Managing for organizational integrity. *Harvard Business Review, 72*(2), 106–117.

Pies, I., Beckmann, M., & Hielscher, S. (2010). Value creation, management competencies, and global corporate citizenship: An ordonomic approach to business ethics in the age of globalization. *Journal of Business Ethics, 94*(2), 265–278.

Pilotta, J. (2013). Confucius and contemporary guanxi. *In International Journal of Area Studies, 8*(1), 128–149.

Reinhardt, F. L. (2005). Environmental protection and the social responsibility of firms: Perspectives from the business literature. In B. L. Hay, R. N. Stavins & R. H. K. Vietor (Eds.), *Environmental protection and the social responsibility of firms: Perspectives from law, economics, and business* (pp. 151–183). Washington, DC: Ressources for the Future.

Rose-Ackerman, S. (1997). The political economy of corruption. In K. A. Elliott (Ed.), *Corruption and the global economy* (pp. 31–60). Washington, DC: Institute for International Economics.

Rüegg-Stürm, J. (2005*). The new St. Gallen management model: Basic categories of an approach to integrated management.* Basingstoke, New York: Palgrave Macmillan.

Sandholtz, W., & Koetzle, W. (2000). Accounting for corruption: Economic structure, democracy, and trade. *International Studies Quarterly, 44*(1), 31–50.

Sanyal, R. (2005). Determinants of bribery in international business: The cultural and economic factors. *Journal of Business Ethics, 59*(1–2), 139–145.

Schein, E. H., & Schein, P. (2017). *Organizational culture and leadership* (5th ed.). Hoboken, NJ: Wiley & Sons.

Shah, A. (2008). Why fighting corruption remains a losing battle. In A. Tarantino (Ed.), *Governance, risk, and compliance handbook: Technology, finance, environmental, and international guidance and best practices* (pp. 133–152). Hoboken, NJ: Wiley & Sons.

Shaw, W. H., & Barry, V. (2016). *Moral issues in business* (13th ed.). Boston, MA: Cengage Learning.

Shleifer, A., & Vishny, R. W. (1993). Corruption. *The Quarterly Journal of Economics, 108*(3), 599–617.

Shleifer, A., & Vishny, R. W. (1998). *The grabbing hand: Government pathologies and their cures.* Cambridge, MA, London: Harvard University Press.

Smart, A., & Hsu, C. L. (2007). Corruption or social capital? Tact and the performance of guanxi in market socialist China. In M. Nuijten & G. Anders (Eds.), *Corruption and the secret of law: A legal anthropological perspective* (pp. 167–190). Aldershot, Burlington, VT: Ashgate.

Steidlmeier, P. (1999). Gift giving, bribery and corruption: Ethical management of business relationships in China. *Journal of Business Ethics, 20*(2), 121–132.

Suchman, M. C. (1995). Managing legitimacy: Strategic and institutional approaches. *Academy of Management Review, 20*(3), 571–610.

Suddaby, R., Bitektine, A., & Haack, P. (2017). Legitimacy. *Academy of Management Annals, 11*(1), 451–478.
Thomson, I., & Boutilier, R. G. (2011). Social license to operate. In P. Darling (Ed.), *SME Mining Engineering Handbook* (3rd ed.) (pp. 1779–1796). Englewood, CO: SME.
Tian, Q. (2008). Perception of business bribery in China: The impact of moral philosophy. *Journal of Business Ethics*, 80(3), 437–445.
Ulrich, P. (2008). *Integrative economic ethics: Foundations of a civilized market economy.* Cambridge et al.: Cambridge University Press.
UN (2003). *United Nations convention against corruption.* https://www.unodc.org/documents/brussels/UN_Convention_Against_Corruption.pdf. Accessed 30 July 2024.
Vandekerckhove, W. (2016). *Whistleblowing and organizational social responsibility: A global assessment.* Abingdon, New York: Routledge.
Verhezen, P. (2009). *Gifts, corruption, philanthropy: The ambiguity of gift practices in business.* Bern: Lang.
Wated, G., & Sanchez, J. I. (2005). The effects of attitudes, subjective norms, attributions, and individualism–collectivism on managers' responses to bribery in organizations: Evidence from a developing nation. *Journal of Business Ethics, 61*(2), 111–127.
Webb, D. K., Tarun, R. W., & Molo, S. F. (2003). *Corporate internal investigations.* New York: Law Journal Press.
Wettstein, F., & Waddock, S. (2005). Voluntary or mandatory: That is (not) the question: Linking corporate citizenship to human rights obligations for business. *Zeitschrift für Wirtschafts- und Unternehmensethik, 6*(3), 304–320.
Wheeler, W. A. (1992). The revival in reverse manufacturing. *Journal of Business Strategy, 13*(4), 8–13.
Wilburn, K. M., & Wilburn, R. (2011). Achieving social license to operate using stakeholder theory. *Journal of International Business Ethics, 4*(2), 3–16.
Wolf, S. (2009). Die Siemens-Korruptionsaffäre – Ein Überblick. In P. Graeff, K. Schröder & S. Wolf (Eds.), *Der Korruptionsfall Siemens* (pp. 9–17). Baden Baden: Nomos.
Wong, Y. H., & Chan, R. Y. K. (1999). Relationship marketing in China: Guanxi, favouritism and adaptation. *Journal of Business Ethics, 22*(2), 107–118.
Wu, X. (2005). Corporate governance and corruption: A cross-country analysis. *Governance, 18*(2), 151–170.
Zimmerli, W. C., & Aßländer, M. S. (2010). Business ethics. In W. C. Zimmerli, K. Richter & M. Holzinger (Eds.), *Corporate ethics and corporate governance* (pp. 37–54). Berlin, Heidelberg: Springer.
Zhou, J. Q., & Peng, M. W. (2012). Does bribery help or hurt firm growth around the world? *Asia Pacific Journal of Management, 29*(4), 907–921.
Zsolnai, L. (2007). Business, ethics and spirituality: Europe–Asia views. *Business Ethics: A European Review, 16*(1), 87–92.

Ethical Leadership: Its Relationship with Social Responsibility

Beatriz Sousa Silva and Carolina Feliciana Machado

Abstract This chapter explores the importance of ethical leadership in the organizational context, highlighting its relationship with social responsibility and other types of leadership. Initially, the concept of ethics and social responsibility is addressed as a basis for understanding the role of ethical leadership, which is distinguished by its emphasis on integrity, transparency, and responsible decision-making. Ethical leadership is compared to other forms of leadership, highlighting its fundamental role in promoting sustainable and socially responsible business practices. The profile of the ethical leader is also discussed, characterized by the ability to inspire trust and positively influence employees through actions that reflect moral values and ethical commitments. By aligning ethical leadership with social responsibility, we can see the positive impact that this type of leadership has on strengthening the reputation of organizations and building a solid and ethical organizational culture. Finally, the importance of ethical leadership for organizations is highlighted, as it fosters a positive work environment, reduces risks related to unethical behavior, and contributes to sustainability and long-term success. Ethical leadership, therefore, emerges as an essential element for organizations that aim not only for financial performance but also for positive social impact.

Keywords Ethics · Leadership · Ethical leadership · Social responsibility · Ethical leader · Organizations

---

B. S. Silva · C. F. Machado (✉)
School of Economics, Management and Political Science, University of Minho, Braga, Portugal
e-mail: carolina@eeg.uminho.pt

C. F. Machado
Interdisciplinary Centre of Social Sciences (CICS.NOVA.UMinho), University of Minho, Braga, Portugal

C. Machado (ed.), *Ethics in Management and Business*, Management and Industrial Engineering, https://doi.org/10.1007/978-3-031-82054-0_7

1 Introduction

Ethical leadership can be considered a crucial phenomenon for organizations, being associated with great benefits. The spread of unethical and socially irresponsible practices by large companies in the 2000s was considered a watershed for the scenario of ethics and social responsibility in the business world due to the urgent need to introduce ethics into organizations and appeal to more responsible practices and conduct.

It is necessary to raise awareness among leaders, managers, employees, and society in general about the importance of these topics, and it is important to realize the real influence that the quality (or lack thereof) of our behaviors has repercussions not only on the lives of the companies where we operate but also in society.

The study of ethical leadership has become increasingly in-depth and is especially pertinent to the uncertain times in which we currently live. Globalization brings with it many changes, and every day, in a business context, quick, assertive, and reactive decisions are necessary. However, this does not invalidate the ability to make correct decisions based on honest values that should not be exceeded.

Therefore, it is necessary to deepen the study even further so that we can obtain more complete information and also for this topic to be disseminated to focus on a future of more ethical and responsible practices, conduct and values, and changing mentalities.

Giving particular focus to the issue of Ethical Leadership, with an emphasis on its relationship with Social Responsibility and its relevance for organizations, this chapter aims, as its main objective, and albeit in a modest way, to serve as a precursor to the discussion of this type of leadership.

2 Ethics and Social Responsibility

Never before has the discussion around Ethics and Social Responsibility taken on such a relevant and necessary position for society. Currently, the evolution is beginning to be quite noticeable, and there are many companies that make reference to these themes in different components of their business, from the integration of ethical and socially responsible values into objectives, values, mission, vision, annual reports and activity plans, even the adoption of codes of ethics and socially responsible policies.

In the digital, technological and globalization era in which we live, the media have the power to disseminate questions about various problems that affect our lives as a community, and problems relating to organizations are no exception. In this way, issues of an ethical nature, and the implications of unethical behaviors, have had greater visibility in the eyes of all of us, leading us to reflect on ethics and morals, on the values by which we govern ourselves and by which those around us and in positions of power are governed.

Attention to ethics and social responsibility in management emerged mainly after the public disclosure of scandals reported in recent decades and after the 2008 financial crisis, which called for an urgent need to change conduct in the business scenario. The most publicized case was the bankruptcy of the company Enron Corporation, in 2001, which became known for the crisis of ethical values in management (Brown and Treviño 2006). We can even say that, "the discourse of ethics became more emphasized the more ethical practices were impoverished" (Rego et al. 2007, p. 42).

It was (and still is) necessary to build, in organizations and business schools, a culture based on values such as honesty, integrity, and excellence, promoting a more ethical management culture. Over the years, ethics training has been included in the academic community, and is considered a necessary pillar for manager training, contributing to increasing the level of ethical practices and raising awareness on this topic (Rego et al. 2007).

But what is ethics, anyway?

The concept of Ethics is widely mentioned, but its definition is neither simple nor consensual. We conceptualize ethics as the discipline of knowledge that reflects on what is right or wrong, moral duty or obligation. It is associated with the judgment of values that govern certain conduct or behavior, being a set of moral principles and values that guide decisions about what we should, or should not, do (Teixeira 2013). It is, therefore, a concept that can be quite subjective, depending on what each of us believes and governs ourselves as a person.

On the other hand, Treviño and Nelson (2011) present a definition of ethics that is more focused on conduct, being defined as "the principles, norms and standards of conduct governing an individual or group" (Treviño and Nelson 2011, p. 17).

Sometimes the concepts of ethics and law can be associated. Many argue that being ethical means complying with the law, and that companies that act in accordance with the law can be considered ethical. However, it is necessary to realize that ethics and law are two distinct concepts. Nothing can guarantee that the law is ethical, and complying with the law does not prove that a particular company or individual is ethical. We have as an example slavery, which was once legalized, or the prohibition of freedom of expression which is still present in totalitarian countries. There needs to be an "ethical reflection", and in these cases, it can be considered that not complying with the law would be ethical (Rego et al. 2007, p. 68).

The concept of Social Responsibility is closely linked to that of Ethics, despite representing and meaning relatively different things. According to Teixeira (2013, p. 369), social responsibility can be considered "the degree to which an organization's managers carry out activities that protect and develop society beyond what is strictly necessary to directly meet the economic and technical interests of the organization". Synthesizing the author's idea, socially responsible companies promote actions that impact society, even if this does not bring any benefit to increasing the company's profit. Organizations must engage in society in an "ethically symbolic" way (Joyner and Payne 2002, p. 292) by "solving problems of public interest" (Etheredge 1999, p. 51) and by recognizing the existence of a need increasing demand for business practices that are ethically responsible (Joyner and Payne 2002).

Investing in promoting ethical and socially responsible conduct has great benefits for organizations. The implementation of a culture and ethical values reinforces and improves the organizational environment, promoting efficiency, productivity, and quality. Furthermore, it strengthens ties and relationships with potential customers, and increases trust with workers, attracting and retaining talent (Su 2014).

The leader becomes a fundamental part of organizations to be able to establish and promote compliance with policies, ethical and socially responsible values. Studies carried out (Bedi et al. 2016; Valentine and Fleichman 2008) show that, by demonstrating their ethical and socially responsible values through communication and their conduct and leadership, leaders will be able to positively influence their employees also to adopt this type of behavior.

3 Leadership

In recent years, Leadership has become crucial for the business world as we live in a more competitive, volatile, and changeable world, with increasingly sudden changes. In the current business context, it is increasingly pertinent and necessary to discuss and investigate leadership, which has become, over the years, a central theme in management and organizational behavior literature (Yukl 1989).

Despite studies, it has been difficult to find a consensual definition, and there is a wide variety of definitions for the concept of leadership.

According to the research team of the Global Leadership and Organizational Behavior Effectiveness (GLOBE) project, leadership is defined as the ability to influence, motivate and allow others to contribute to the success of the organizations in which they operate (House et al. 1999).

In turn, Yukl (2010, p. 7), goes a little deeper into this concept, stating that leadership can be defined more broadly, as the "process of influencing others to understand and agree about what needs to be done", facilitating "individual and collective efforts to accomplish shared objectives", making "current work" simpler, and ensuring that everyone is "prepared to meet future challenges".

A study was conducted by Cunha et al. (2009), as cited in Cunha et al. (2014), to try to understand the relevance of leadership in organizations. The authors conclude that the perception that the individuals involved have about the leader can influence the entire organization. If the perception about the leader is positive, the organization will benefit from it, whereas if it is a negative perception, it will bring disadvantages to the organization. That said, the authors consider leaders as "organizational translators", since "good leaders" intensify positive situations and neutralize negative ones (p. 158). Therefore, we can see that, in fact, leadership has a great influence on organizations and is a determining factor in their ability to achieve success.

Effective leaders (and those who contribute positively to organizational performance and success) have a set of personality traits associated with them. Hiregoudar and Vani (2018), Kirkpatrick and Locke (1991), and Yukl (2010) identify a series of key characteristics for effective leadership, of which it is necessary to highlight integrity and honesty, trust, commitment, responsibility, empathy, resilience and emotional intelligence, transparency, humility, altruism and the ability to empower others.

These characteristics are also associated with a specific type of leadership: ethical leadership.

4 Ethical Leadership

Ethical leadership has aroused the interest of researchers, especially after the release of ethical scandals that exposed the behaviors of leaders of large organizations to everyone's eyes.

Just like the basic concepts of leadership and ethics, the concept of Ethical Leadership can be difficult to define. There have been many studies that have analyzed and attempted to define ethical leadership, the way leaders should act and its implications.

To better understand this concept, it is important to mention a study by Treviño et al. (2000) where leaders are described under two dimensions: The moral person and the moral manager.

The 'moral person' dimension represents the 'ethical' part of the concept of ethical leadership and is crucial for the leader to be able to develop a reputation as an ethical leader, appropriate in the eyes of his collaborators, as they will associate him with individual traits, such as honesty, trust and integrity. Therefore, we can say that this dimension refers to the qualities of the leader as a person, their principles, values, altruistic character, making decisions based on justice and equity. However, being a moral person is not, in itself, enough for someone to be considered an ethical leader. To complement this portrait, they must be able to act as 'moral managers', proactively promoting ethical conduct and values and demonstrating the organization's ethical culture. Having both dimensions, the ethical leader becomes the organization's model of ethical conduct, and must regularly communicate ethical standards with their employees and use systems that can ensure that these standards are being followed (Treviño and Nelson 2011).

In the absence of any of these dimensions, we are not faced with an ethical leader, and one of the following options may occur (Treviño and Nelson 2011, p. 158): "unethical leadership", "hypocritical leadership" or "ethically neutral leadership". Leaders may be perceived by their employees as being neither moral people nor moral managers, and are therefore considered unethical. These leaders may display traits of disrespect, violence, and dishonesty, for example. On the other hand, if the leader is a strong moral manager but a weak moral person, he is a hypocritical leader, as he tries to instill ethical conduct in others, without being able to follow it himself. We can say that, in addition to being hypocrites, this type of leader is also dishonest. Finally, if the leader is considered a strong moral person but is a weak moral manager, then we can assert that he will be an ethically neutral leader as he fails in terms of leadership, communication, open dialogue, and the promotion and application of systems that verify ethical conduct in the organization (Treviño et al. 2000).

Another perspective, which corroborates and completes the previous definition, is that of Brown et al. (2005). Based on the social perspective of learning, ethical leadership is defined as a "demonstration of normatively appropriate conduct through personal actions and interpersonal relationships, and the promotion of such conduct to followers through two-way communication, reinforcement, and decision-making" (p. 120).

Ethical leaders are seen as role models, based on certain characteristics that make them legitimate and credible, such as honesty and loyalty. Furthermore, they make decisions based on principles, revealing prudence, compassion, and empathy for others. They not only draw attention to ethics, but also highlight it by talking openly about it, promoting a climate of trust. It needs to be able to establish ethical standards and be able to reward those who follow ethical standards or discipline those who do not follow these guidelines, contributing to its learning (Brown et al. 2005).

4.1 Ethical Leadership and Other Associated Forms of Leadership

Ethical leadership has characteristics that are similar to other theoretical conceptions of leadership, namely transformational, transactional, spiritual, and authentic leadership (Brown and Treviño 2006).

Brown and Treviño (2006) state that ethical and transformational leadership have some similarities since, in both leadership styles, leaders care about others, are seen as models to be followed, and act in accordance with their values and principles.

However, there are some noticeable differences between both conceptions. Ethical leadership is mainly characterized by moral, ethical, and fair behavior as a way of influencing those around it. In transformational leadership, this is not so noticeable. In transformational leadership, leaders stimulate critical and innovative thinking in their followers to become aware of problems and help them to recognize their own principles and values, encouraging them intellectually to adopt a vision of the future and change (Brown and Treviño 2006; Den Hartog 2015).

Another notable difference between the two is the fact that the ethical leader is seen as a moral person and a moral manager. In contrast, transformational leaders can exhibit more altruistic and authentic or more egocentric intentions (called pseudo-transformational) (Bass and Steidlemeier 1999). Clearly, due to the explicit focus on morality, the authentic form of transformational leadership shares more similarities with ethical leadership, with the pseudotransformational form identifying more with unethical leadership (Brown and Treviño 2006).

Transactional leadership has a point closely associated with ethical leadership, especially when it comes to the act of rewarding subordinates when they practice and adopt a certain type of measure. Ethical leaders use reward or punishment to shape the behavior of their followers (Brown and Treviño 2006). However, the focus in transactional leadership is on getting followers to achieve expected performance and how they can facilitate that performance. On the contrary, in ethical leadership, despite also adopting reward and punishment, the focus goes far beyond that, wanting to actually affect the ethical awareness and behavior of subordinates (Den Hartog 2015).

In relation to spiritual leadership, Brown and Treviño (2006) also point out similarities and differences in relation to the concept of ethical leadership. Spiritual leadership focuses on motivating followers to develop a "sense of spiritual survival" (Fry 2003, p. 711). Furthermore, these types of leaders are visionary and spiritual, a characteristic not associated with ethical leadership. The convergence with ethical leadership arises because leaders are also considered to be integrity, honest, humble, trustworthy, treating others ethically, compassionately and with respect (Reave 2005).

Finally, authentic leadership can be considered a "root structure" that could "incorporate charismatic, transformational, or integral and/or ethical leadership" (Luthans and Avolio 2003). This type of leadership also converges with ethical leadership in terms of the individual characteristics of its leaders. Both authentic leaders and ethical leaders have ethical principles (they act in accordance with them) and are people-oriented. However, authentic leadership stands out for its authenticity and self-awareness, principles that are not fundamental to the concept of ethical leadership, which focuses more on ethics and morality, care and concern for others (Brown and Treviño 2006; Den Hartog 2015).

5 How Is an Ethical Leader Characterized?

According to Ajzen (1987), individual differences and characteristics can be very useful in understanding someone's tendencies to make certain decisions of an ethical nature or to adopt certain behaviors. By associating characteristics with our behaviors and actions, we can understand more about ethical decision-making and ethical behavior.

In addition to the characteristics that we associate with ethical leadership that have been identified and described in the previous points, such as loyalty, trust, and integrity, among others, it is possible to characterize the ethical leader according to their personality traits.

The most publicized personality model is The Big Five, and after its creation, it is possible to conceptualize and measure personality characteristics more effectively.

Its popularity is mainly due to the abundant empirical support that has been collected. This model encompasses five essential characteristics: conscientiousness, extroversion, agreeableness, emotional stability, and openness to experience (Cunha et al. 2014).

Conscientiousness is associated with responsibility, obedience, determination, and perseverance (Brown and Treviño 2006). This characteristic describes individuals who are self-disciplined and resistant to impulses, being highly related to professional performance (Cunha et al. 2014).

Extroversion can be associated with energy and assertiveness and describes individuals who are affectionate, sociable and feel good about themselves. Extroversion can also be called "positive affectivity" (Cunha et al. 2014, p. 45).

Agreeableness is highly related to altruism, trust, and kindness and is associated with individuals who are able to establish good interpersonal relationships (Cunha et al. 2014).

Neuroticism corresponds to the tendency to experience negative emotional states and to experience little emotional stability. The most associated characteristics are hostility, impulsivity and stress.

Lastly, openness to experience refers to the degree to which an individual is willing to try new experiences and adapt to change. This type of individuals can be associated with imagination, curiosity, originality and creativity (Brown and Treviño 2006).

Brown and Treviño (2006) propose that conscientiousness and agreeableness are personality characteristics positively related to ethical leadership, not proposing any relationship with openness to experience and extroversion, since these personality traits are more related to charisma, a characteristic that does not define ethical leadership.

Therefore, the authors consider that kindness may be the personality trait that most relates to this type of leadership, since, by definition, these types of leaders are altruistic, attentive and concerned about those around them (Brown and Treviño 2006).

Conscientiousness also plays an important role in defining the characteristics of an ethical leader, representing self-control, reliability, responsibility and trust. Individuals with this type of characteristic tend to think carefully before acting, always taking into account their moral obligations and responsibilities, which is extremely important for leaders who want to be perceived as being ethical (Kalshoven et al. 2011). To achieve this, leaders establish clear and well-defined ethical standards (which must be applied to their lives) so that they can be seen as ethical leaders. Synthesizing the authors' point of view, conscientious individuals are more likely to be perceived as being ethical leaders (Kalshoven et al. 2011).

Finally, with regard to neuroticism, this may be a characteristic that is also not related to ethical leadership since it is not even a characteristic that forms a leader (Hogan et al. 1994). Walumbwa and Shaubroeck (2009) did not find any type of relationship between ethical leadership and neuroticism since ethical leaders represent exactly the opposite of what an individual with this characteristic is, since it is associated with hostility, fear, and anxiety (Den Hartog 2015).

Therefore, it can be stated that neuroticism is not linked to the characteristics of an ethical leader, and it is expected that these leaders do not express negative emotional states or behave impulsively. Instead, ethical leaders are associated with high levels of emotional stability (Kalshoven et al. 2011), being able to maintain positive interpersonal relationships with their subordinates based on trust and respect.

6 Ethical Leadership and the Relationship with Social Responsibility

Social responsibility has been an increasingly important topic for organizations, and we already know that it is highly related to ethics.

Because ethical leaders are associated with characteristics such as concern for others, trust, responsibility, reliability, altruism, moral awareness, and ethics, they may also be associated with making more socially responsible decisions.

Winter and Barenbaum (1985) developed a measure capable of assessing the level of social responsibility of individuals, which focuses on whether the person assessed shows an inner obligation to do what is right, whether they take responsibility for themselves and others, whether they are capable demonstrate self-control, are aware of the consequences of their actions and are trustworthy (De Hoogh and Den Hartog 2008).

These five elements are seen as constituents of social responsibility, and De Hoogh and Den Hartog (2008) propose that they are related to ethical leadership, mainly in relation to morality and equity, the clarification of roles and the sharing of powers. The authors' study shows that leaders scoring high in terms of social responsibility were associated with high levels of ethical leadership. The social responsibility of leaders was positively associated with ethical leadership, as well as with morality and equity, characteristics adjacent to this type of leadership (De Hoogh and Den Hartog 2009).

Therefore, we can affirm that ethical leaders, in addition to making choices based on fair principles, acting with integrity and being trustworthy, and acting based on justice and integrity, when looking at the interests of those around them, can also be considered socially responsible leaders, and there is even an internal obligation to do so. Leaders who care about social responsibility behave ethically and care about the consequences of their own actions on others.

7 Importance of Ethical Leadership for Organizations

Ethical leadership has been identified as a great advantage for both organizations and their employees, bringing great benefits to both parties.

Firstly, according to the social learning perspective, ethical leaders are perceived by their subordinates as being credible and motivating and, therefore, are considered models to be followed, presenting correct behaviors that could be "imitated" by subordinates. Furthermore, ethical leaders, through their conduct and values, are able to communicate and talk about ethics openly. Therefore, they positively influence their subordinates also to adopt this type of ethical conduct.

Brown and Treviño (2006) propose that ethical leaders influence ethical conduct in organizations and proliferate it throughout the rest of the organization, almost assuming a position of "ethical guidance". His followers know that the leader "will be holding them accountable for their decisions and will use rewards and discipline to do so" (Brown and Treviño 2006, p. 607). Therefore, subordinates are more likely to focus on their conduct, and the ethical implications their decisions have, leading them to make more ethical decisions. Furthermore, ethical leadership is also associated with the reduction of counterproductive behaviors on the part of employees, as they replicate the actions of their leader, which makes the standards of conduct very clear, as well as the consequences of violating such standards (Brown et al. 2005).

Ethical leadership style also brings with it many benefits for employees which in turn can also be considered benefits for the organization. According to Bedi et al. (2016), ethical leadership stimulates employee effort and dedication, the adoption of ethical behaviors, job satisfaction, psychological well-being, organizational commitment and involvement and identification with the organization. When leaders treat employees fairly and communicate the importance of ethics, employees are less likely to leave the organization, become involved in conflicts, and adopt counterproductive behaviors (Bedi et al. 2016).

These leaders maintain relationships based on transparency and honesty with their subordinates, worrying about their physical and psychological well-being. Therefore, the relationship between leader and subordinate is a strong relationship of trust, in which the leader also stimulates the "voice" of the subordinate, asking for their opinion when making decisions, thus also reinforcing communication (Brown el al. 2005).

In this way, we can consider that the proliferation of ethical conduct becomes essential for organizations and that the role of ethical leadership, through the leader, is crucial for them to be able to reinforce and improve the entire organizational environment.

8 Final Remarks

With this chapter, we can see that the practice of ethics is absolutely essential, and the adoption of ethical and responsible leadership styles becomes a great advantage for organizations. Corporate scandals and their harmful and perverse effects have brought an urgent need for change, which has occurred over the years.

Ethical decision-making requires that one pays attention and think about the consequences for all parties involved, from employees, managers, customers to society. There are inalienable principles and rights that can never be overcome. It is important that learning ethics is present in everyone's training, especially leaders, as it becomes essential for the sustainable development of organizations and life in society.

Ethical leadership is important for developing ethical conduct and moving towards more ethical and responsible organizational cultures. Ethical management and social responsibility "are not done by decree" (Rego et al. 2007, p. 19) and it is necessary that there is a moral and ethical motivation, that there is awareness, and that respectful relationships are created, based on communication, so that a good organizational and work climate proliferates.

In this sense, the role of the ethical leader is crucial so that values and ethical conduct can be disseminated, perceived and understood. By being a model to be followed, and by speaking openly about these topics, the ethical leader promotes ethics throughout the organization, resulting in several benefits for them: greater productivity, better working conditions, greater well-being among workers, good organizational climate, among many others listed throughout this chapter.

In short, it is important to emphasize that we must move towards more ethical organizational cultures since their inception. The incorporation of ethical values in organizations and their practice, by everyone, is essential, since ethical decision-making avoids unfair, dishonest and inhumane situations. Encouraging the learning of basic values such as transparency, honesty, justice, are essential for both the business world and personal life. It is not enough to be good "moral managers", we must also be "moral people".

References

Ajzen, I. (1987). Attitudes, traits, and actions: Dispositional prediction of behavior in personality and social psychology. Advances in Experimental Social Psychology, 20(1), 1–63.

Bass, M., & Steidlmeier, P. (1999). Ethics, character, and authentic transformational leadership. The Leadership Quarterly, 10, 181–217.

Bedi, A., Alpaslan, C., & Green, S. (2016). A Meta-analytic review of ethical leadership outcomes and moderators. Journal of Business Ethics, 139(3), 517–536.

Brown, M., & Treviño, L. (2006). Ethical leadership: A review and future directions. The Leadership Quarterly, 17(6), 595–616.

Brown, M., Treviño, L., & Harrison, D. (2005). Ethical leadership: A social learning perspective for construct development and testing. Organizational Behavior and Human Decision Process, 97(2), 117–134.

Cunha, M., Rego, A., Cunha, R., Cardoso, C., & Neves, P. (2014). Manual de Comportamento Organizacional e Gestão (7ª ed.). Lisboa: Editora RH.

De Hoogh, A., & Den Hartog, D. (2008). Ethical and despotic leadership, relationships with leader's social responsibility, top management team effectiveness and subordinates' optimis: A multi-method study. The Leadership Quarterly, 19(3), 297–311.

De Hoogh, A., & Den Hartog, D. (2009). Ethical leadership: The socially responsible use of power. In D. Tjosvold & B. Wisse (Eds.), Power and Interdependence in Organizations (Cambridge Companions to Management, (pp. 338–354). Cambridge: Cambridge University Press.

Den Hartog, D. (2015). Ethical Leadership. Annual Review of Organizational Psychology and Organizational Behaviour, 2(1), 409–434.

Etheredge, J. (1999). The percieved role of ethics and social responsibility: an alternative scale structure. Journal of Business Ethics, 18(1), 51–64.

Fry, L. (2003). Toward a theory of spiritual leadership. The Leadership Quarterly, 14, 693–727.

Hiregoudar, S., & Vani, G. (2018). Qualities of Good Leadership for Effective Organisation. International Journal of Research and Scientific Innovation, 5(4), 336–338.

Hogan, R., Curphy, G. J., & Hogan, J. (1994). What we know about leadership: Effectiveness and personality. Amercian Psychologist, 49(6), 493–504.

House, R., Hanges, P., Quintanilha, S., Dortman, P., Javidan, M., Dickson, M., & Grupta, V. (1999). Cultural influences on leadership and organizations: Projects GLOBE. Advances in Global Leadership, 1(2), 171–233.
Joyner, B., & Payne, D. (2002). Evolution and Implementation: A study of values, business ethics and corporate social responsibility. Journal of Business Ethics, 41(4), 297–311.
Kalshoven, K., Den Hartog, D., & De Hoogh, A. (2011). Ethical Leader Behaviour and Big Five Factors of Personality. Journal of Business Ethics, 100(2), 349–366.
Kirkpatrick, S., & Locke, E. (1991). Leadership: do traits matter? Academy of Management Executive, 5(2), 48–60.
Luthans, F., & Avolio, B. (2003). Authentic leadership: A positive development approach. In K. S. Cameron, J. E. Dutton, & R.E. Quinn (Eds.), Positive Organizational Scholarship—Foundations of a new discipline (pp. 241–258). San Francisco: Berrett–Koehler.
Reave, L. (2005). Spiritual values and practices related to leadership effectiveness. The Leadership Quarterly, 16, 655–687.
Rego, A., Cunha, M., Costa, N., Gonçalves, H., & Cardoso, C. (2007). Gestão Ética e Socialmente Responsável (2ª ed.). Lisboa: Editora RH.
Su, H. Y. (2014). Business ethics and development of intellectual capital. Journal of Business Ethics, 119(1), 87–98.
Teixeira, S. (2013). Gestão das Organizações. Lisboa: Escolar Editora.
Treviño, L., & Nelson, K. (2011). Managing Business Ethics: Straight talk about how to do it right (5th ed.). New York: John Wiley & Sons.
Treviño, L., Hartman, L., & Brown, M. (2000). Moral Person and Moral Manager: How executives develop a reputation for ethical leadership. California Review Management, 42(4), 128–142.
Valentine, S., & Fleischman, G. (2008). Professional Ethical Standards, Corporate Social Responsibility, and the Percived Role of Ethics and Social Responsibility. Journal of Business Ethics, 82(3), 657–666.
Walumbwa, O., & Schaubroeck, J. (2009). Leader personality traits and employee voice behaviour: Mediating roles of ethical leadership and work group psychological safety. Journal of Applied Psychology, 94(5), 1275–1286.
Winter, D., & Barenbaum, B. (1985). Responsibility and the power motive in women and men. Journal of Personality, 53, 335–355.
Yukl, G. (1989). Managerial Leadership: A Reviwe of Theory and Research. Journal of Management, 15(2), 251–289.
Yukl, G. (2010). Leadership in Organizations (7th ed.). Upper Saddle River: Prentice Hall.

Sexual Harassment in an Organizational Context: Ethical Decision-Making Process

Ana João Gonçalves Ferreira and Carolina Feliciana Machado

Abstract This chapter examines the problem of sexual harassment in organizations from the perspective of ethical decision-making, addressing the challenges and responsibilities of organizations and individuals when dealing with this sensitive issue, seeking to highlight the application of Thomas Jones' contingency model and its multidimensional component—Moral Intensity. Initially, the chapter characterizes sexual harassment as a complex ethical problem, whose consequences go beyond the individual impact, affecting organizational culture and the work environment. Next, key ethical decision-making process models that offer frameworks for evaluating and responding to ethical dilemmas in the business context are explored. These models guide managers and employees in analyzing the consequences, duties, and virtues involved in decisions about sexual harassment. Emphasis is given to the approach adopted by Jones, which provides a more comprehensive and adaptable understanding, emphasizing how different situations and contexts influence the ethical response. The dimensions of Moral Intensity are explored, which allow the severity of ethical dilemmas to be assessed based on factors such as magnitude of consequences, proximity, and probability of effect, facilitating understanding of the ethical and emotional impacts involved. The study contributes to reinforcing the importance of a robust and responsible ethical approach in organizations, emphasizing the need for preventive policies and support for victims to promote a safe and ethical work environment.

Keywords Sexual harassment · Organizational context · Ethical problem · Moral intensity · Ethical decision-making process

---

A. J. G. Ferreira · C. F. Machado (✉)
School of Economics, Management and Political Science, University of Minho, Braga, Portugal
e-mail: carolina@eeg.uminho.pt

C. F. Machado
Interdisciplinary Centre of Social Sciences (CICS.NOVA.UMinho), University of Minho, Braga, Portugal

C. Machado (ed.), *Ethics in Management and Business*, Management and Industrial Engineering, https://doi.org/10.1007/978-3-031-82054-0_8

1 Introduction

The main objective of this chapter is to analyze the ethical problem of sexual harassment in an organizational context.

More specifically, the individual's decision-making process and the role of organizations in relation to this type of behavior will be analyzed. This study is based on Thomas M. Jones' contingency model and the multidimensional componentmoral intensity. magnitude of consequence

The chapter will begin with a brief description and analysis of the concept of sexual harassment and why it is considered an ethical problem. Next, a small review of the literature will be made on the different theories and dimensions of the ethical decision-making process and, finally, as mentioned previously, the application of the six variables that explain moral intensity, namely, the magnitude of consequences, social consensus, effect of actions, temporal immediacy, proximity between actors and concentration of responsibility.

Through this analysis we will be able to understand the dynamics of the decision-making process; what leads the individual to act or not to act on a certain behavior (sexual harassment) in an organizational context; the factors that determine the different levels of moral intensity and the importance of the role of organizations in defining individuals' behaviors.

2 Ethical Problem

Sexual harassment is a social problem associated with discrimination and gender inequality. It is also related to a legal problem, as there are laws and sanctions that punish this type of behavior. It still must also be considered as an ethical problem that is part of the relationships that are established within organizations (Crain and Heischmidt 1995).

According to Ferrell and Fraedrich (1991), as cited in Crain and Heischmidt (1995, p. 302), this concept is considered "the behavior of organization members toward customers, suppliers, subordinates, superiors, peers, and others".

As such, it becomes necessary to define the phenomenon of sexual harassment. According to McCann (2005), the concept emerged in the 1970s in the United States of America, as a result of the civil rights movement period, in which women (a substantially affected party) demanded that sexual harassment be recognized as discriminatory behavior by law. Inspired by this movement, several countries, international organizations, and companies have developed and introduced the concept of sexual harassment through policies, practices and codes of conduct and ethics.

As cited by McCann (2005) and in accordance with the European Commission Directive (2002/73/EC, Article2(2)), sexual harassment takes the form of "[...] unwanted verbal, non-verbal or physical conduct of a sexual nature occurs, with the purpose or effect of violating the dignity of a person, in particular when creating an intimidating, hostile, degrading, humiliating or offensive environment" (p. 10).

In an organizational context, the North American Equal Employment Opportunity Commission (EEOC), as cited by O'Leary-Kelly and Bowes-Sperry (2001, p. 75), defines sexual harassment as "unwelcome sexual advances, requests for sexual favors, and other verbal or physical conduct of a sexual nature...when (1) submission to such conduct is made either explicitly or implicitly a term or condition of an individual's employment, (2) submission to or rejection of such conduct by an individual is used as the basis for employment decisions affecting such individual, or (3) such conduct has the purpose or effect of unreasonably interfering with an individual's work performance or creating an intimidating, hostile or offensive work environment".

The definition presented by the North American Commission initially refers to the Latin expression *Quid pro quo*, in which two or more parties establish an agreement in which there is a reciprocal exchange of favors, such as a salary increase or career promotion, in exchange for favors of a sexual nature. The exposed concept also addresses the idea of a hostile and harmful work environment for the victim, which encompasses a series of behaviors, such as insults based on gender, derogatory comments about the victim, innuendo, and display of sexually suggestive or explicit material between others.

It is important to note that the previously mentioned behaviors are not always understood as sexual harassment. If some are easily recognized as offensive and harmful to workers, such as physical aggression, other behaviors are dependent on the context in which the action occurred, the relationships between individuals in the organization, and the purpose and tone in which it was uttered, such as sexual jokes.

According to the American Heritage Dictionary (1982), as cited in O'Leary-Kelly and Bowes-Sperry (2001, p. 73), the concept of "ethics" is defined as "a principle of right or good conduct; a system of moral principles or values". The reflection and assessment that individuals make about sexual harassment behaviors result from a system of values, norms, and codes present in society that determines what is or is not morally acceptable. According to Jones (1991), "an ethical decision occurs when an individual resolves a moral issue through a choice that is both legal and morally acceptable to the larger community" (p. 367).

As such, we can say that the way we choose to treat others in the workplace results from an ethical-moral decision. The same applies to the decision-making process of organizations in situations of sexual harassment.

In the study carried out by Bowes-Sperry and Powell (1999), the authors state that "observers of social-sexual behavior" (p. 779) when they recognize such behavior as an ethical problem, there is a tendency to adopt ethical behavior—intention and action of intervening. In contrast, many workers view the sexual harassment laws and policies of the organizations they work for as ineffective and an invasion of privacy. Many simply do not consider this issue serious, resulting in part from the inaction of organizations that do little to change their employees' perceptions of this issue (O'Leary-Kelly and Bowes-Sperry 2001).

Therefore, how can we explain this problem of inaction on the part of individuals and organizations when they do not recognize the importance or moral imperative of the problem? To understand this question, one of the answers found is associated with the literature on the decision-making process of individuals, namely Thomas M. Jones' contingency model and the multidimensional concept of moral intensity.

Authors Bowes-Sperry and Powell (1999) concluded that there was a positive relationship between levels of moral intensity and individuals' recognition of sexual harassment behaviors as being an ethical issue on which they should act. Hence the importance of people's conduct to solve this problem, as organizations cannot rely exclusively on the role of victims to report cases of harassment, especially because many of them are afraid of losing their jobs, generating reprisals against the harasser and above all, they need to maintain their reputation and status in the work group, which means that many cases go unreported (Bowes-Sperry and O'Leary-Kelly 2005).

It is important to mention that the act of harassment is not always observed directly, as in *quid pro quo* situations. However, when we talk about the creation and/or perpetuation of a hostile and harassing work environment, the feeling is general and recognized by the majority.

3 Models of the Ethical Decision-Making Process

As previously mentioned, in the literature, the decision-making process of individuals was addressed in the Jones Contingency Model through the Moral Intensity component (on which this chapter focuses) but also other authors have presented a set of variables that explain or influence the decisions and course of action of individuals.

Rest (1986) cited by Jones (1991), presented a model consisting of four components that explain the cognitive process that is part of ethical decision-making. According to the author, the moral agent recognizes the ethical dilemma (problem recognition), which in turn thinks about the action he should take (ethical judgment). Then, the agent reveals the intention and the way in which he will proceed (ethical intention), which translates into the last phase, ethical behavior, which takes the form of action. The author argues that the different components are independent and that the success of one stage does not imply the completion of the next stage.

Kolberg's (1976) theory of moral development, cited by Jones (1991) points to an explanation for the previously mentioned idea, as the author argues that the way individuals behave depends on their level of moral development. The three subdivided levels (pre-conventional, conventional, and post-conventional) represent the various stages of cognitive development in which the individual finds himself. When progression to the next level occurs, there is an evolution in moral reasoning.

This passage to the next level does not always occur, and even if a person has conventional cognitive development (they maintain order and live according to rules that are socially accepted) it does not mean that their behavior follows ethical principles beyond those that are determined by current laws and regulations.

Trevino (1986) came to complement Rest's model (1986) and determines that the ethical judgment that individuals make is dependent on personality characteristics, such as the internal locus of control, which translates into the perception that the individual has about the control it has, that is, the outcome of a situation is or is not the product of its action and also the strength of the ego. This scale measures the level of individuals' convictions and self-regulation (Ribeiro 2000). The author also cites the influence of situational factors, such as the context in which the work is performed, the organizational culture, and the type of work performed.

Authors such as Ferrell and Gresham (1985), mentioned by Jones (1991), associated other elements to the ethical decision-making process, such as the values, knowledge, attitudes, and intentions of individuals, as well as organizational aspects such as the existence (or not) of codes of conduct, rules and a system of compensation and punishment.

Hunt and Vitell (1986), as cited by Jones (1991), also add personal experiences as a component that affects the perception of the existence of an ethical problem, the associated alternatives, and its consequences. Together with the duties, principles, and regulatory norms (deontological norms) that govern individuals in an organization, they determine the ethical judgment formed, which in turn directly affects the person's intention and behavior.

Dubinsky and Loken (1989) used Fishbein and Ajzen's (1975) theory of reasoned action as a basis to explain the ethical decision-making process. This theory assumes that human beings are rational beings, who analyze information systematically, always considering the implications of their actions before acting, which does not mean that the behavior adopted is ethical or morally accepted.

According to Jones (1991), the previously presented models contributed to the understanding and deepening of the study of the ethical decision-making process, however, the components and characteristics presented reduce the individual's decision-making and behavior in the organization, as being identical for all ethical problems. That is, "people will decide and behave in the same manner whether the issue is the theft of a few supplies from the organization or the release of a dangerous product to the market" (p. 371). The author mentions that the models presented do not consider the characteristics of the ethical problem in question.

To overcome this limitation, Jones (1991) developed the notion of moral intensity, a multidimensional concept that analyzes the level of intensity (moral imperative) on the ethical problem in the decision-making process, namely the higher the level of intensity, the greater the probability of the individual adopting ethical behavior.

The concept of moral intensity becomes relevant or functions as a variable that influences the problem of sexual harassment in organizations, because as previously mentioned, it is not an issue treated with the seriousness and importance it requires. Organizations have a fundamental role here, as they are responsible and can contribute to increasing the moral intensity of the problem, through defining the behaviors of individuals within organizations, namely developing a culture of prevention and reporting for this type of behaviors, with the creation of codes of conduct and ethics and training actions on harassment in the workplace, allowing employees to make more ethical choices.

Through the different dimensions of the concept of moral intensity it is possible to explain and understand the individual's decision-making process and the role of organizations with regard to this issue.

4 Dimensions of Moral Intensity

The **magnitude of the consequences** refers to the impact (negative or positive) of the ethical decision. Based on what was said by O'Leary-Kelly and Bowes-Sperry (2001), generally, they are "[...] acts that have potential to cause greater numbers of people to suffer, or that may cause more severe suffering, are considered to have a higher level of moral intensity" (p. 78).

Men and women perceive sexually harassing behavior in different ways; in particular, while a man interprets a certain behavior or attitude as being sexual in nature, a woman sees it as being merely friendly. On the other hand, men consider more friendly attitudes from the opposite sex to be flattering, while most women consider it an insult. According to Riger (1991), "[...] women are more likely to see as harassment more subtle behavior such as sexual teasing or looks or gestures" (p. 499). The author also mentions that there is a natural tendency for females to identify a greater number of sexual harassment behaviors compared to males.

The different interpretations lead us to think about the different intensity levels for both genders. If a man interprets a woman's behavior as an invitation of a sexual nature, there is a greater likelihood that his actions and/or responses will follow the same pattern. Not considering their behavior as causing suffering or harmful to the victim. The moral intensity of the magnitude of the consequences of his behavior is low and the decision-making process does not assume ethical principles.

Another aspect that influences individuals' behavior is related to the use of humor in interactions between them, particularly associated with sexual harassment behaviors. Humor works as a deterrent that takes away the seriousness and validity of actions, reducing attitudes or words to a simple joke, often of a sexist nature (O'Leary-Kelly and Bowes-Sperry 2001). This assumption makes it difficult for workers to analyze the behavior, justifying their actions as just a joke, thus limiting the actors (direct or indirect) to the ability to perceive the magnitude and negative consequences of the behavior.

The magnitude of the consequences depends on its severity and the number of people it affects; thus, "[...] individuals may regard sexual harassment as a relatively benign issue because they believe that it affects only the individual who is the target of harassment" (O'Leary-Kelly and Bowes-Sperry 2001, p. 80). Different authors claim that this correlation does not represent reality and speak of the negative effects of sexual harassment in organizations, such as low levels of productivity and performance and negative psychological impact on all members of the organization (Crain and Heischmidt 1995).

The level of **social consensus** refers to the approval that exists on a given issue (Jones 1991). According to the author, when there is a high level of consensus, there is a greater probability for individuals to act ethically (standard of socially acceptable behavior). However, the ambiguity that exists around the concept of sexual harassment determines the low level of moral intensity. This ambiguity manifests itself in the conduct of organizations. On the one hand, we have organizations that only consider sexual harassment as explicit acts of a sexual nature. In contrast, other organizations consider harassment beyond behaviors of a sexual nature, which are also related to gender discrimination.

Individuals in the decision-making process also take into account the reaction of the harassment victim's behavior (reception or rejection). This idea is directly related to one of the prejudices that exists regarding sexual harassment behavior, incitement. These consider that the victim provokes such behavior, through the way they express themselves or dress. Finally, other factors, such as professional and personal experience also determine the decision-making process (O'Leary-Kelly and Bowes-Sperry 2001).

Organizations have a leading role in clearly defining and covering the concept of sexual harassment, hence the importance of developing codes of conduct and ethics that act as a guiding element in the behavior of workers within organizations. The individual is aware of the organization's position, knows how to act, the consequences, and the impact that its actions have.

The **effect of actions**, according to Jones (1991), concerns the degree of probability of a certain action with moral content occurring. For the author, an action that is more likely to cause harm makes the individual recognize the moral intensity of that act and be more cautious before acting.

Related to this dimension is the behavior of the victim of harassment; most of the time, the victim does not react as expected, that is, there is a tendency to minimize the problem, such as ignoring the harassing behavior and/or relativizing this same behavior (O'Leary-Kelly and Bowes-Sperry 2001). This type of reaction does not create the need for the individual to intervene, as the victim appears not to be affected by the situation.

Another element that determines ethical behavior is the interpretation and reception of the individual's behavior by the organization to which he belongs. Organizations with low levels of tolerance for sexual harassment behaviors encourage individuals through periodic meetings and anti-harassment policies.

In these organizations, the associated risk is lower, and the moral intensity is greater, as the individual is aware that their actions will be taken seriously and will have positive effects. On the other hand, the cost–benefit analysis that individuals carry out determines a tendency for work relationships to deteriorate and possible retaliation, particularly non-career progression, isolation, or even dismissal, especially if the aggressor occupies a position of decision-making or power (Bowes-Sperry and O'Leary-Kelly 2005).

This component is directly related to the dimension of **temporal immediacy**, which concerns the temporal space between the present and the beginning of the consequences of a given act (Jones 1991). Actions are recognized as having greater moral intensity when the result occurs within a short space of time. The passage of time minimizes the magnitude of the consequences and limits the actions of individuals.

Victims of sexual harassment often do not immediately recognize the harassment they have suffered (there is doubt and questioning). It is often only recognized when the behavior is repeated, and this involves a wider period of time, as well as the negative impact of the traumatic event (Paetzold and O'Leary-Kelly 1993). In an organizational context, victims fear that the organization will discredit the situation and hope that it will naturally ease over time. Considering these elements, we can state that the moral intensity in the decision-making process will be low, as its action may not have immediate results.

The **proximity** dimension between the actor and the target also affects the moral intensity of the behavior. In light of what was written by Jones (1991), the similarity between individuals (cultural, social, and psychological) helps to determine the moral intensity of their actions. This idea is part of the social categorization principles theory by authors Hogg and Terry (2000). For both, individuals are part of groups with certain characteristics that share the same values and principles. If the harasser is male, there is a tendency for the same gender group to evaluate his behavior more positively. Individuals feel obligated to help and support members who are part of their clan. In organizations with traditionally male environments, there is a greater propensity to experience sexual harassment towards the minority, the female sex.

The structure and hierarchy within organizations also determine the behavior of individuals. The power relations that exist within organizations point to the interpretation or identification of harassing behaviors in vertical relationships and not so much in horizontal relationships between colleagues.

Finally, there are organizations with a "sexualized" work environment, which allow flirtatious behavior, sexual jokes, and aggressive language. According to the authors Victor and Cullen (1988), as cited in O'Leary-Kelly and Bowes-Sperry, the role of the organization involves creating an ethical-moral organizational culture where it is defined "what ethical conflicts are considered, the process by which such conflicts are resolved, and the characteristics of their resolution" (2001, p. 75).

In the **concentration of responsibility** dimension by Jones (1991), the author states that when a small group of people suffers a significant number of negative consequences, it produces a more concentrated moral effect compared to a larger group in which the consequences are less impactful.

The psychological effect associated with the hostile environment of harassment in organizations is not calculated to be a high or significant cost to the entire organization. Latané and Darley (1970) point to the connection of this dimension to the effect of diffusion of responsibility, that is, when several individuals witness a certain behavior, they place or assume that the other is responsible for intervening or acting on the problem.

This idea is related to the concept of influence of the social environment, that is, the reaction that others have to a situation determines the group's behavior. If other group members do not reveal discomfort with a particular observed behavior, it is likely that others will not make the decision to intervene. Latané and Darley (1970), as cited in Bowes-Sperry and O'Leary-Kelly (2005), argue that "the presence of other people serves to inhibit the impulse to help" (p. 295).

Paetzold and O'Leary-Kelly (1993) state that the relationships that are established in a work context make individuals assume that there may be a pre-existing personal relationship, in which they may or may not consider it correct to intervene, even if the sexually harassing behavior among managers and colleagues occur. In other cases, the individual does not intervene because they do not trust their judgment about the action; their interpretation of the sexual harassment behavior may not be correct.

There are also organizations that have workers and/or departments whose role involves dealing with this type of case, namely the human resources department. Here there is a transfer of responsibility to this group, considering these elements of the organization to be more competent and able to act (Bowes-Sperry and O'Leary-Kelly 2005).

These factors highlight the ambiguity that exists regarding the concept of sexual harassment. The failure to define rules and forms of action by organizations is at the origin of constraints in the decision-making process and in the moral intensity of actions. Once again, the organization's role involves creating awareness among its employees that everyone is responsible and has a duty to report and prevent these behaviors from being perpetuated in the workplace and, therefore, in society in general.

5 Final Considerations

Sexual harassment is a phenomenon that is part of the reality of organizations, as it occurs in work relationships. It is a concept that extends beyond physical contact and the exchange of favors of a sexual nature; it is also a hostile and intimidating work environment for the worker.

Sexual harassment can be considered an ethical problem since the analysis that individuals make of sexual harassment behaviors results from a system of values, norms, and codes embedded in society that determines what is or is not morally acceptable.

It was understood that there are a set of variables that influence the ethical decision-making process, namely the recognition of the ethical dilemma, the individual's cognitive development, personality characteristics, and situational factors. However, Jones (1991) pointed to the need to create a component that truly explains what leads an individual to adopt a certain ethical behavior, for which the author presented the multidimensional concept—moral intensity. This concept becomes relevant to explain the problem of sexual harassment in an organizational context, as organizations play a fundamental role in defining individuals' behaviors in the workplace.

With the analysis of the six dimensions of moral intensity, it was concluded that there are different interpretations and other aspects such as the use of humor as a deterrent. Added to this is the ambiguity that exists regarding the concept, as well as the idea of incitement, professional and personal experience, and the influence of the social environment, which limit the decision-making process. Another aspect that is important to mention is the relativization of the problem and the way in which the organization views the phenomenon.

Taking these elements into account, it is necessary to create organizations with low levels of tolerance for sexual harassment. Create an awareness that breaks with fears and prejudices, through the definition of codes of conduct and ethics that give the organization a clear and objective view of the problem to workers. Know how to act, the consequences and impact that their actions have, combined with anti-harassment training and practices, such as case reporting systems and delimitation of workers' rights and duties, are examples. In conclusion, the existence of enlightened and informed employees facilitates the ethical decision-making process and determines greater levels of moral intensity.

References

Bowes-Sperry, L., & O'Leary-Kelly, M. A. (2005). To act or not to act: The dilemma faced by sexual harassment observers. Academy of Management Review. 30(2), 288–306. https://doi.org/10.5465/amr.2005.16387886

Bowes-Sperry, L., & Powell, G. N. (1999). Observers' reactions to social-sexual behavior at work: An ethical decision making perspective. Journal of Management, 25(6), 779–802. https://doi.org/10.1177/014920639902500601

Crain, A. K., Heischmidt, A. K. (1995). Implementing Business Ethics: Sexual Harassment. Journal of Business Ethics. 14(4), 299–308. https://doi.org/10.1007/BF00871900

Dubinsky, J. A., Loken, B. (1989). Analyzing ethical decision making in marketing. Journal of Business Research. 19(2), 83–107. https://doi.org/10.1016/0148-2963(89)90001-5

Hogg, A. M., & Terry, J. D. (2000). Social Identity and Self-Categorization Processes in Organizational Contexts. Academy of Management Review. 25(1), 121–140. https://doi.org/10.2307/259266

Jones, M. T. (1991). Ethical decision making by individuals in organizations: An issue-contingent model. Academy of Management review. 16(2), 366–395. https://doi.org/10.5465/amr.1991.4278958

Latané, B. & Darley, J. M. (1970). The unresponsive bystander: why doesn't he help? Appleton-Century Crofts

McCann, D. (2005) Sexual harassment at work: national and international responses. Project Report. International Labour Organization, Geneva. http://www.ilo.org/travail/whatwedo/publications/WCMSTRAVAILPUB2/lang−−en/index.htm

O'Leary-Kelly, M. A., & Bowes-Sperry, L. (2001). Sexual harassment as unethical behavior: the role of moral intensity. Human Resource Management Review. 11(1–2), 73–92. https://doi.org/10.1016/S1053-4822(00)00041-3

Paetzold, R. L., & O'Learly-Kelly, A. M. (1993). Continuing violations and hostile environment sexual harassment: When is enough, enough. American Business Law Journal. 31 (3), 365–395. https://doi.org/10.1111/j.1744-1714.1993.tb00685.x

Ribeiro, C. (2000). Em torno do conceito Locus de Controlo. Máthesis, (9), 297–314. https://doi.org/10.34632/mathesis.2000.3854

Riger, S. (1991). Gender Dilemmas in Sexual Harassment Policies and Procedures. American Psychologist Association. 46 (5), 497–505. https://doi.org/10.1037/0003-066X.46.5.497
Trevino, K. L. (1986). Ethical Decision Making in Organizations: A Person-Situation Interactionist Model. Academy of Management review. 11(3), 601–617. https://doi.org/10.2307/258313

Index

C. Machado (ed.), *Ethics in Management and Business*, Management and Industrial Engineering, https://doi.org/10.1007/978-3-031-82054-0

M

N

O

The manufacturer's authorised representative in the EU is Springer
Nature Customer Service Centre GmbH, Europaplatz 3, 69115 Heidelberg,
Germany. If you have any concerns regarding our products, please
contact ProductSafety@springernature.com

Printed and bound by CPI Group (UK) Ltd, Croydon, CR0 4YY
30/09/2026
02225599-0001